REAL Science Odyssey

Read ⁘ Explore ⁘ Absorb ⁘ Learn

Chemistry
Level 1

written by Blair H. Lee, M.S.

ISBN: 978-0-9798496-3-3

The publisher and author have made every attempt to state precautions and ensure that all activities and labs described in this book are safe when conducted as instructed, but assume no responsibility for any damage to property or person caused or sustained while performing labs and activities in this or any RSO course. Parents and teachers should supervise all lab activities and take all necessary precautions to keep themselves, their children, and their students safe.

WHAT'S INSIDE THIS BOOK?

*Denotes lab or activity

Pandia PRESS

An Introduction to RSO Chemistry

Atoms, molecules, chemical reactions, the periodic table—concepts intimidating enough to make a person weak in the knees! Young children can learn chemistry! Students as young as seven years old can successfully complete this course even though many of the concepts taught in this book are those usually reserved for high school. Don't worry, this book is written with the science novice in mind. Even if you have never been taught chemistry or don't know an electron from a proton, you will find yourself learning right alongside your child, and looking like a pro the whole time! You and your child will never look at a common piece of matter (like your shoe) the same again, without thinking about the billions and billions of molecules and atoms that make it up.

This course is not a collection of random labs, meant to entertain but with little real science. Nor is it a long progression of fill-in-the-blanks and trivial facts to be memorized and forgotten. Instead, this book is rich in vital concepts that will lay a firm foundation for studying chemistry in later years. This book is intended to be used from start to finish, like a math book. Concepts are taught through clever repetition and engaging labs, building upon one another and introducing science vocabulary and age-appropriate math gently and in context. There is particular emphasis placed on scientific method throughout this course. Students will learn how to speculate, hypothesize, experiment, observe, interpret, and conclude, just like real scientists.

A good chunk of this course is devoted to learning about the periodic table and studying several important elements on the table. Any serious study of chemistry must include the periodic table. It is the chemist's alphabet and it is as vital to chemistry as letters are to literature. When students are in a chemistry class later in life, they will have the background to advance further; they will not fear the table.

There is evidence that science is best learned when taught as a single subject and not lumped together with multiple mini-subjects taught in the span of a school year. That said, all the sciences do interrelate. Chemistry is the study of atoms, what they make, and how they interact. Since everything is made from atoms, the argument can be made that chemistry is the central science. Unfortunately, it is the science least likely to be taught before high school. The problem with that is, if you try to understand a scientific concept with any degree of depth, you had better know some chemistry. I have thought for some time now that educators start teaching chemistry too late. It is a fascinating subject, but there are new facts and new vocabulary, and you are learning about things that are the size of an atom—way too small to see! With chemistry, more than any other science, you often have to memorize a fact and take scientists' word for what is happening. The time to do that is not when someone is in high school or college; it is when a child is in the beginning stage of learning. This is the ideal time for absorbing facts, memorizing content, and filling a child's mind with the knowledge that creates a firm foundation for later years.

While creating this book, a lot of thought was given to teaching to the three main learning styles—visual, auditory, and tactile/kinesthetic. Science, by its very nature, lends itself to teaching to all three styles. With that in mind, each lesson in this book has visual, auditory, and tactile components. For visual learners, there are diagrams and charts that students help create and visualize. There are illustrations coupled with the text. There are crossword puzzles to help learn vocabulary words. The experiments have been carefully selected and designed to support the concepts taught in the text. Therefore, the experiments themselves are powerful visual aids.

For auditory learners, those who learn best through listening, there is the text itself, which has been designed to be read to students. There is a poem in each unit of this book to help reinforce vocabulary and key concepts. There are questions peppered throughout the text and within the experiments to help these students think through and hear the answers, thereby cementing the concepts being taught in an auditory fashion.

For tactile/kinesthetic learners, those who learn through moving, doing, and actively exploring the physical world, this book has been designed for them as well. There are activities that use movement to teach concepts. There are drawings and models to make. There are puzzles for students to put together. There are diagrams and charts to create. There are hands-on labs to perform, which directly teach to the concepts.

Science is a creative and thought-provoking endeavor. It is the one academic subject where you are supposed to move around and get your hands dirty. With such an emphasis on test scores these days, it seems breadth and depth in academics has been forgotten. It is time to bring science back as a core part of curriculum, engaging young people's minds with the stuff they really find interesting. Let's get started!

THE UNIQUE PAGES IN THIS BOOK

For My Notebook Pages

1. All the student pages have a boxed outline around the material presented. That way it is easy to identify what is for the child and what is for the parent or teacher.
2. The For My Notebook (FMN) pages are the lesson pages that present the majority of new material to the student. They are intended to be read aloud. Some students, who are good readers, may want to read the FMN pages aloud themselves to the parent or class. However orchestrated, these pages are intended to be read aloud and not silently, to encourage discussion and questions.
3. New vocabulary words are underlined. You will notice that many of the vocabulary words are not presented with a classic dictionary definition. Instead, the explanation is given in context so it is "felt" rather than memorized. Formal definitions for the vocabulary words are offered in the back of the book.

Lab Sheets

1. The lab sheets are those pages that the student writes on. They also have a boxed outline because they are intended for the student, not the parent/teacher, to complete.
2. The lab sheets not only reinforce the material presented in the FMN pages, but they are also the vehicle through which this course reinforces and formalizes scientific method. On the lab sheets, students will be making hypotheses based on questions formed during the lesson. Students record observations and lab results, and make conclusions based on those results. They will also practice sketching details of their lab experiences, an important process that reinforces observation skills.
3. If you are working with a student who isn't writing yet, then have him dictate the information to be written on the lab sheets. If your student is unable to draw (meaning physically incapable; I'm not referring to artistic abilities), then have him describe in detail his observations as you create them on the lab sheet.

The Instructor Pages

1. The instructor pages contain the supply lists for the labs or activities and procedure instructions.
2. These pages are written for the parent/teacher, but the procedure is often written as if for the student. For example, "Complete the hypothesis portion of the lab sheet," is instruction for the student, not the parent.
3. Most instruction pages include a prompt to read aloud to students. A great deal of course instruction is found in these prompts. If you dislike prompts, then be sure to present the information in your own words.

Poem Pages and Crossword Puzzles

1. Each unit contains a poem and a crossword puzzle that are intended to help reinforce vocabulary and key concepts.
2. The poems can be used as you wish—recite, memorize, transcribe, illustrate. Concepts learned to verse are learned more quickly and not as easily forgotten.
3. The crossword puzzles review the vocabulary presented in each unit.

What's the Big Idea?

Whenever you study a subject, there are main ideas and details to learn. It's true, that in science, there is a lot of new material to discover. This outline gives you the big ideas that your child should get from each unit and the small stuff that is an added bonus. If you and your child are timid scientists, just have fun as you try to learn the big ideas. If you and your child have a strong science background, work on learning the small stuff as well as the big ideas. There are many challenging words in this course that are used because they are the right words, and after hearing them over and over, they will "sink in." They are not here for your child to memorize the first time around. Use difficult words and science concepts gently, not with force, and your child will enjoy their science experience.

BI = BIG IDEA **SS = SMALL STUFF**

UNIT 1 - WHAT IS CHEMISTRY?

WHAT IS CHEMISTRY?

BI = All things are made of chemicals.
Chemistry is the science that studies chemicals.
A chemist is a person who studies how chemicals interact.
A hypothesis is your best guess about the outcome of an experiment.

SS = Chemists use two types of tests to tell different things apart—physical tests and chemical tests.
Most chemicals are benign.

UNIT 2 - STARTING SMALL

THE ATOM

BI = Atoms are extremely small.
Everything is made of atoms.
Atoms move.

SS = A Greek philosopher named Democritus talked about the concept of atoms 2,400 years ago.
Temperature affects how fast atoms move and the rate of diffusion.
A scanning-tunneling microscope is needed to see atoms.

PARTS!

BI = The three parts of an atom are protons, neutrons, and electrons.
Protons and neutrons are found in the nucleus.
Electrons are found orbiting the nucleus in energy levels.
All protons, neutrons, and electrons are like every other proton, neutron, and electron.

SS = Protons and electrons are charged particles.
Atoms are mostly empty space.

TYPES!

BI = Each type of atom has a unique name.
The thing that makes one type of atom different from another type of atom is the number of protons that the atom has in its nucleus.
The number of electrons a neutral atom has is the same as the number of protons in its nucleus.
An element is a group of the same type of atoms.

SS = A maximum of two electrons go in the first energy level. A maximum of eight electrons go in the second energy level.
When going from one type of atom to the next, one and only one proton is added to the nucleus.
Most atoms have more than one energy level.

UNIT 3 - THE CHEMIST'S ALPHABET DEFINED

THE ALPHABET

BI = The periodic table is a chart that has the names of all the elements on it.
Everything is made from these elements.
Elements are on the periodic table in abbreviated forms, called symbols.

SS = Dmitri Mendeleev invented the periodic table.
Some symbols come from names of the elements that are still used today.
Some symbols come from earlier names used for the element.

ATOMIC NUMBERS

BI = The atomic number is found on your periodic table above the symbol for the element.
The atomic number equals the number of protons in the nucleus of that type of atom.
The number of electrons in an atom's elemental form equals the atomic number.

SS = If something is an element, all parts of it will react the same in a chemical test.
When going from one element to the next one on the periodic table, one proton is added to the nucleus.

MASSIVE MATTERS

BI = The atomic mass of an element equals the number of protons + the number of neutrons.
Mass is the same anywhere in the universe. Weight is affected by gravity.

SS = Things that take up the same amount of space can have different masses.
If the number of the same type of atoms stays constant, the liquid state weighs the same as the solid state.
a.m.u. = atomic mass unit. It is the unit used to measure the atomic mass of an atom.

WHY DO THEY CALL IT THE PERIODIC TABLE ANYWAY?

BI = The rows of the periodic table are called periods.
There are seven periods on the periodic table.
The periodic table is read from left to right and down, like a book.

SS = The period an element is in equals the number of energy levels it has.

WE ARE FAMILY

BI = The columns of the periodic table are called groups.
Elements in the same group on the periodic table share traits with each other.
There are 18 groups on the periodic table.

SS = Elements in the same group have the same number of electrons in their outer energy level.

IT'S ELEMENTARY

BI = The top three rows (periods) on the periodic table have 18 commonly occurring elements.

SS = 92 of the 118 elements are naturally occurring, the rest are manufactured.

UNIT 4 - THE CHEMIST'S ALPHABET APPLIED

Unit 4 is a step-by-step, group-by-group, explanation of how the elements on the Periodic Table are arranged, based on their structure.

BI = One and only one proton is added to the nucleus of an element when going from one element to the next on the periodic table.
The number of protons in an element's nucleus equals its atomic number.
The atomic mass of an element equals the atomic number plus the number of neutrons.
The total number of electrons for elements on the periodic table equals the number of protons for that element.
All elements in the same group have the same number of electrons in their outer energy level.

SS = The number of neutrons does not increase consistently but does increase overall.

HE LIKES NACHOS – GROUP 1

BI = Introduction to the elements hydrogen, lithium, and sodium.

SS = Things are more buoyant in NaCl + water than in water alone.

BE MGNIFICENT - GROUP 2

BI = Introduction to the elements beryllium and magnesium.

SS = A solution = solute + solvent.

Water is the universal solvent.

BUMBLEBEES ALIGHT – GROUP 13

BI = Introduction to the elements boron and aluminum.

CONSTANTLY SILLY – GROUP 14

BI = Introduction to the elements carbon and silicon.

SS = Carbon browns when heated.

Sugar is made from carbon, hydrogen, and oxygen.

NICE PENGUINS – GROUP 15

BI = Introduction to the elements nitrogen and phosphorus.

OBNOXIOUS SEAGULLS – GROUP 16

BI = Introduction to the elements oxygen and sulfur.

SS = When hydrogen peroxide bubbles it is turning into water and O_2.

Hydrogen peroxide bubbles on cuts because of catalase in your blood, not germs in the cut.

FREQUENTLY CLEVER – GROUP 17

BI = Introduction to the elements fluorine and chlorine.

SS = Fluorine bonds to calcium in teeth making teeth stronger.

HE NEVER ARGUES – GROUP 18

BI = Introduction to the elements helium, neon, and argon.

SS = Gas shrinks (has a smaller volume) when it is cold.

UNIT 5 - MOLECULES RULE

PUTTING IT ALL TOGETHER

BI = Atoms bond (link) together and make molecules.

The bonds form when atoms share electrons from their outer energy levels.

SS = A compound is a group of all the same type of molecules.

A mixture is a group of different types of molecules.

MOLECULAR FORMULAS ARE USEFUL

BI = Molecular formulas tell the amount and type of atoms present in a molecule.

To write a molecular formula, write the symbol for the element and the number of atoms present.

SS = Capillary action in plants is a result of the interaction between water molecules and the cellulose molecules in the plant.

DRAWING LESSONS

BI = There is a special method for drawing molecules called Electron Dot Structures.

When drawing molecules, place the atoms so they are sharing electrons on the side where there is a single electron.

SS = Electron Dot Structures pair electrons from different atoms using only electrons in the atoms' outer energy levels.

The rate of capillary action can be affected by how "sticky" water finds molecules it encounters.

UNIT 6 - WHAT'S THE MATTER?

MATTER: AN INTRODUCTION

BI = Protons + neutrons + electrons make atoms. Atoms group together to make molecules.

Molecules group together to make matter.

SS = Sometimes when two different types of molecules get together, they switch atoms around and make different molecules. When this happens, it is called a chemical reaction.

When chemical reactions happen, atoms cannot be created or destroyed.

THE STATES OF MATTER

BI = Matter is anything that takes up space and has mass (weight).
Matter comes in three states—solid, liquid, and gas.
The three states of matter are differentiated by the physical properties of definite shape and volume.

SS = Water is found in all three states—ice, liquid water, and steam—over the normal temperature range.

LET'S GET TO THE POINT

BI = The special name for the point where matter goes from solid to a liquid = melting point.
The special name for the point where matter goes from liquid to a solid = freezing point.
The special name for the point where matter goes from gas to a liquid = condensation point.
The special name for the point where matter goes from liquid to a gas = boiling point.

SS = The temperature of the freezing point = the temperature of the melting point.
The temperature of the boiling point = the temperature of the condensation point.
The melting point of water = freezing point = 32˚F (0˚C)
Boiling point of water = condensation point = 212˚F (100˚C)
Some matter does not fall completely into one of the three defined states.

SOLIDS ARE DENSE

BI = The molecules in a solid are close together.
The molecules in a solid do not move much, they vibrate slightly and have a fixed position relative to each other.
Solids are more dense than gases.
Density = the amount of stuff in a given space.

SS = Solids have a definite shape and a definite volume.
Solids are more dense than liquids, with the important exception of water.
Density is affected by the amount and type of molecules or elements present.

MOLECULES STICK TOGETHER

BI = Molecules stick together by sharing electrons.

SS = Molecules in solids are close together and don't move much.
Molecules in liquids are farther apart and move more.
Molecules in gases are not very close together—they move a lot.

WHAT MAKES A LIQUID A LIQUID?

BI = The molecules in a liquid are close together but not as close as in a solid.
The molecules in a liquid move around but not as much as in a gas.
Liquids have a definite volume, but not a definite shape.

SS = Different types of liquids can have different densities.

THE FRIENDSHIP OF OXYGEN AND HYDROGEN

BI = Water is an important and unique compound. It is essential to life, as we know it.

SS = Hydrogen bonds hold water molecules together.
Because ice is less dense than liquid water, the same number of molecules in ice has a greater volume than those molecules in liquid water.
Water can be separated into hydrogen and oxygen gases through a process called electrolysis.

WHAT MAKES A GAS A GAS?

BI = The molecules in a gas have a lot of space between them.
The molecules in a gas are fluid, move very fast, and mix freely.
Gases do not have a definite volume or a definite shape.
Gases are less dense than solids and liquids.

SS = Because gases are less dense than liquids, they will float to the top and out of a liquid.

THE AIR YOU BREATHE

BI = Oxygen is essential to life.

Air takes up space.

SS = Air is 78% nitrogen and 21% oxygen,

Oxygen is cycled between people, animals, and plants.

Air is important for regulating the temperature on Earth.

UNIT 7 - REACTIONS IN ACTION

CHANGES

BI = Chemical change = chemical reaction

In a chemical reaction, the molecules at the beginning are not the same as those at the end. The same atoms are there, but they are bonded differently.

When there is a physical change, the molecules present do not change.

In a chemical reaction, the starting molecules are called reactants.

In a chemical reaction, the ending molecules are called products.

SS = Evidence of chemical change = temperature change, the formation of bubbles, or a solid that is forming.

Chemical reactions are written as equations.

Exothermic reactions release heat.

Endothermic reactions absorb heat.

SOME LIKE IT SOUR, SOME DON'T

BI = Acids taste sour.

Bases taste bitter and feel slippery.

Indicators tell if a solution is an acid, a base, or neutral.

SS = In acids, the hydrogen atom comes off in a solution and leaves an electron. This is called a dissociation reaction.

In bases, the oxygen and hydrogen atoms come off in solution and take electrons. This is also called a dissociation reaction.

HYDROGEN AND OXYGEN AND HYDROGEN MAKE WATER

BI = acid + base = water + salt = an acid-base reaction

SS = H without its electron (from an acid) and OH with an electron (from a base) makes H_2O.

If an acid and a base solution are mixed, and the resulting solution is neutral, it is called a neutralization reaction.

pHUNNY pHRIENDS

BI = The pH scale is used to measure the strength of an acid or a base.

SS = The pH scale measures from 0 to 14.

7 is neutral.

Less than 7 is an acid.

Greater than 7 is a base.

BUILDING TEETH

BI = In precipitation reactions, different liquids are mixed together and one of the products formed is a solid.

SS = Precipitation reactions are the type of chemical reactions used to make bones, teeth, shells, and coral reefs.

COMBUSTION ACTION

BI = Combustion reactions must have oxygen gas as one of its reactants.

SS = Combustion reactions are exothermic.

If oxygen gas is taken away, the fire goes out and the reaction stops.

Lab Supply List (see page 2 for RSO supply kit information)

Items are listed by unit in the order in which they are first needed. + indicates that an item will be needed for later labs also. The amounts listed are totals for the entire course. Most items are common household items.
* indicates that the item requires some explanation. Ordering hints or explanations are given on the next page.

UNIT	EQUIPMENT / MATERIAL	AMOUNT
1+	Teaspoon	3
1+	Clear glass	3
1	Dish	3
1+	Kitchen towel	1
1	Box of cereal	1
1+	Confectionery sugar	2 ¾ cups
1+	Baking powder	1 cup
1	Baby powder	½ cup
1+	White vinegar	5 cups
1+	Cooking oil (vegetable oil)	2 cups
1+	Food coloring (blue)	1 bottle
1	Dish soap	1
2+	Balloons	6
2	Almond extract	1 tablespoon
2+	Water (tap)	1 -2 gallons
2+	Cinnamon	2 tablespoons
2	Lemon, orange, or peppermint extract	1 tablespoon
2+	Magnifying glass (hand lens)	1
2	½ teaspoon measuring spoon	1
2	Permanent marker	1
2+	Crayons or colored pencils	1 box
2	Construction Paper (8 ½" x 11")	1
2+	School glue (Elmer's glue)	2 bottles
2	Inflated balloon	1
2	Mirror	1
2	Carpet or fabric sofa	1
2	Aluminum foil	
2	Pencil lead or a sharpened pencil	1
2	Mini marshmallows (all one color)	1 bag
2+	Regular-size marshmallows (two colors)	2 bags + 4
2+	Blank sheets of paper 8 ½" x 11"	20
2+	Kitchen scale that measures grams	1
2+	Scissors	1
3+	Lemon juice	1 cup
3+	Paintbrush	1
3+	Heat source, such as a lamp	1
3+	Pot holder	2
3+	Milk	1 ½ cup
3+	Measuring cup	various sizes
3+	Stapler	1
3+	Internet access	
3+	Chemistry books/encyclopedias	
3	Frozen water in a baggie	1 cup
3	Brown sugar	1 cup
3	Grapes	1 cup
3	Grape juice	1 cup

UNIT	EQUIPMENT / MATERIAL	AMOUNT
3+	Re-sealable baggies	10
3+	Peanut butter	1 small jar
3	Powdered milk	1 cup
3	Honey	¼ cup
3+	Wooden spoon	1
3+	One-gallon freezer baggie	4
3+	Mixing bowl	1
3+	Tablespoon	1
3+	Fork	1
3+	Table salt	1 ½ cups
3	Salt substitute (potassium chloride)	¼ cup
3	Cooked potato	1
3	Crushed ice	2 cups
3+	Small cup	2
3	1/8 teaspoon measuring spoon	1
3+	Thermometer (science or kitchen-type*)	2
3+	Eggs	7
3+	Flour	3 cups
3+	Granulated sugar	5 cups
3	Cinnamon sugar (optional)	1 tablespoon
3	Muffin pan	1
3	Muffin cup liners	12
3	Whisk	1
3+	Oven and stove top with a timer	1
4	12" x 12" card stock (different colors)	10 sheets
4	Glue stick	1
4+	Distilled water	10 cups
4	Epsom salt	3 tablespoons
4	Black construction paper 8 ½" x 11"	1
4	20 Mule Team Borax*	2 teaspoons
4	Grated styrofoam or polystyrene beads*	1 cup
4+	Food coloring	1 bottle
4	Small cake pan	1
4	Graham crackers	2
4+	Baking sheet / cookie sheet	1
4	Brown colored pencil	1
4	Egg whites	3
4	Parchment paper	
4	Shortening or nonstick spray	
4	Electric mixer	1
4+	Pot	1
4+	Knife	1
4+	Hydrogen peroxide	1 bottle
4	Raw potato	1
4+	Sink or work bucket	1
4	Bleach	½ cup

UNIT	EQUIPMENT / MATERIAL	AMOUNT
4	Eyedropper	1
4	Fluoridated toothpaste	4.6 oz.
4	Toothbrush	1
4	Colored nail polish	
4	Plastic wrap	
4	Helium-filled balloons	2
4	Flexible cloth tape measure	1
4+	Freezer	1
5	8 ½" x 11" card stock	1
5	Bag of gumdrops (assorted colors)	1
5	Toothpicks	1 box
5	Pepper	½ teaspoon
5	Lettuce	1 head
5	Tomato	1
5	Carrot	1
5	Salad bowl	1
5	Stalk of celery	1
5+	Chalkboard or dry erase board	1
5	WHITE paper towel with NO designs on it	1
5	Shallow dishes	3
5	Cotton swabs (Q-tips)	1
5	Ruler	1
5+	Tape	
6	LEGO blocks	10 - 15
6	Ice cubes	2 cups
6	Rock	1
6	Drinking straw (optional)	1
6	Crushed ice	1 cup
6	Jello	1 box
6	Cold water	1 cup
6+	Hot water	2 cups
6	Mayonnaise	1 tablespoon
6+	Refrigerator	1
6	Plate	1
6	Medium-sized box or container	1
6	Stuffed animals	10
6	Wash tub, bathtub, or sink filled with water	1
6+	Orange	2
6	Small (1- to 2-cup size) plastic container	1
6	Marbles	10 - 20
6	Assortment of waterproof solids	8
6+	Quart-size glass jar with lid	1
6	6-inch long piece of rough string or yarn	1
6	Clean metal washer or a Lifesaver candy	1
6	Funnel (optional)	1
6	Corn syrup	¼ cup
6	Clear 2-cup container	1
6	Empty plastic soda bottles	2
6	Pan for under soda bottles	1

UNIT	EQUIPMENT / MATERIAL	AMOUNT
6	6-inch insulated copper wires	2
6	9-volt battery	1
6	Can of soda	1
6	Popping corn	1 cup
6	Empty 2-liter soda bottle	1
7	Alka-Seltzer tablet	1
7+	Matches or lighter	1
7	Kool-Aid	2 packs
7+	Pitcher with pour spout	1
7	Bubble-blowing solution and wand	1
7	Cast-iron skillet or fire-proof container	1
7	Yeast	1 teaspoon
7	Clear measuring cup	1
7+	Baking soda	11 teaspoon
7	Head of red cabbage	1
7	Strainer	1
7	Large 8-cup container with a cover	1
7	White coffee filters	2
7	Rubber gloves	1 pair
7	Ammonia	2 teaspoons
7	Laundry detergent	1 teaspoon
7	7-Up	1 teaspoon
7	Clear plastic disposable cups	10
7	Newspaper	1
7	Red, blue, and white watercolor paint	1 T of each
7	Paint palette	1
7	Lemon	1
7	Grapefruit	1
7	Lime	1
7	Cherry Tomato	1
7	V-8 juice	1 cup
7	Chalk	6 pieces
7	Different size glass jars	3
7	Roasting pan	1
7	Sand	5 cups
7	Votive candles	3
7	Stopwatch	1
7	Dollar bill	1
7	Tongs	1
7	91% Rubbing alcohol	½ cup

***Ordering Hints:**

Thermometer: The liquid in good science thermometer goes down on its own, unlike the liquid in a medical thermometer, which must be shaken down. Some kitchen thermometers will work, but most do not go low enough. You will need a thermometer that will go down to 30° F or 0° C.

20 Mule Team Borax: A laundry booster found next to laundry detergents in grocery stores.

Polystyrene beads: Used for stuffing stuffed animals. Found in some craft stores in the sewing section. Grated Styrofoam is an excellent substitute.

Suggested Weekly Schedule

The following schedule is suggested for those wishing to complete this course in a 36-week school year, teaching science twice a week. General supplies needed for each week are listed. Refer to the lesson or supply list for specifics on supplies including quantities. FMN indicates For My Notebook lesson pages.

Week	Day	Lesson / Lab	Supplies Needed for the Week	Dates / Notes
1	Day 1	What Is Chemistry? (FMN) Be a Chemical Detective	Box of cereal, Confectionery sugar, Baking powder, Baby powder, Water, Vinegar, Cooking oil, Blue food coloring, Teaspoons, Pour container, Clear glasses, Dishes, Kitchen towel, Sink, Dish soap	
	Day 2	Telling Things Apart		
2	Day 1	The Atom (FMN) Are Atoms Small?	Balloons, Almond extract, Water, Cinnamon, Lemon or peppermint extract, Magnifying glass, Measuring spoon, Permanent marker, Water at three temperatures, Food coloring, Clear glasses, Thermometer, Colored pencils	
	Day 2	Do Atoms Move?		
3	Day 1	Parts! Let's Be Positive	Crayons or colored pencils, Construction paper, Glue, Inflated balloon, Wall, Mirror, Carpet or fabric sofa, Large work surface, Aluminum foil, Pencil lead, Mini marshmallows, Regular-size marshmallows, Blank sheets of paper, Kitchen scale, Scissors	
	Day 2	Types! (FMN) The First Ten		
4	Day 1	The Alphabet (FMN) My Periodic Table	Lemon juice, Paintbrush or cotton swab, Sunlight or lamp, Periodic table, Pot holder	
	Day 2	Chemical Symbol Match My Favorite Element		
5	Day 1	Atomic Numbers (FMN) Is Milk an Element?	Milk, Glass, Lemon juice, Measuring cup, Scissors, Colored pencils, Stapler	
	Day 2	Flipbook		
6	Day 1	Massive Matters (FMN) My Favorite Element Explored	Internet access and/or chemistry books, Colored pencils or crayons, Water (frozen), Liquid water, Powdered Sugar, Brown sugar, Grapes, Grape juice, Kitchen scale, Measuring cup, Sealable baggies	
	Day 2	Which Weighs More?		
7	Day 1	Why Do They Call It the Periodic Table Anyway? (FMN) Periodic Table Worksheet	Periodic table, Peanut butter, Honey, Powdered milk, Sealable baggie, Mixing bowl, Measuring cup, Wooden spoon	
	Day 2	Periodic Play Dough		
8	Day 1	We are Family (FMN) The Friendship of Beryllium and Boron	Periodic table, Table salt, Salt substitute (potassium chloride), Oil, Vinegar, Cooked potato (optional), Water, Crushed ice, Glass, Cups, Measuring spoon, Towel, Science or kitchen thermometer	
	Day 2	Prove It!		
9	Day 1	It's Elementary (FMN) Twenty Questions	Periodic table, Flour, Salt, Cinnamon, Eggs, Sugar, Milk, Vegetable oil, Baking powder, Cinnamon sugar (optional), Muffin pan, Muffin cup liners, Mixing bowls, Whisk, Measuring cup, Measuring spoon, Oven, Hot pads	
	Day 2	Eating Hockey Pucks		
10	Day 1	Element Book He Likes Nachos - Group 1 (FMN)	Periodic table, 12″ x 12″ card stock, Scissors, Glue stick, Stapler, Art supplies (markers, colored pencils, crayons), Eggs, Tall clear glasses, Distilled water, Salt, Tablespoon, Stirrer	
	Day 2	The Incredible Floating Egg Element Book Group 1		
11	Day 1	Be Mgnificent - Group 2 (FMN) Crystal Creation	Epsom salt, Black construction paper, Warm water, Magnifying glass, Measuring cup, Tablespoon, Cake pan, Scissors, Periodic table, Scissors, Glue, Art supplies (markers, colored pencils, crayons)	
	Day 2	Element Book Group 2		
12	Day 1	Bumblebees Alight - Group 13 (FMN) The Slime That Ate Slovenia	20 Mule Team Borax, Water, White school glue, Grated Styrofoam or polystyrene beads, Food coloring (optional), Freezer baggie, Mixing bowl, Measuring cup, Teaspoon, Tablespoon, Fork, Periodic table, Scissors, Glue, Art supplies (markers, colored pencils, crayons)	
	Day 2	Element Book Group 13		

Week	Day	Lesson / Lab	Supplies Needed for the Week	Dates / Notes
13	Day 1	Constantly Silly - Group 14 (FMN) S'more Carbon	Large marshmallows, Graham crackers, Baking sheet, Oven, Hot pad, Brown colored pencil, Periodic table, Scissors, Glue, Art supplies (markers, colored pencils, crayons)	
	Day 2	Element Book Group 14		
14	Day 1	Nice Penguins - Group 15 (FMN) Eating Air	Egg whites, Confectionery sugar, Parchment paper, Nonstick spray, Flour, Cookie sheet, Mixing bowl, Fork, Spoon, Electric mixer, Oven, Hot pads, Magnifying glass, Periodic table, Scissors, Glue, Art supplies (markers, colored pencils, crayons)	
	Day 2	Element Book Group 15		
15	Day 1	Obnoxious Seagulls - Group 16 (FMN) That's Not My Egg You're Cooking, Is It?	Eggs, Clear glass, Heat source, Pan, Water, Timer, Salt, Knife, Hydrogen peroxide, Raw potato, Sink, Periodic table, Scissors, Glue, Art supplies (markers, colored pencils, crayons)	
	Day 2	Bubble Trouble Element Book Group 16		
16	Day 1	Frequently Clever - Group 17 (FMN) Dancing Drops	Bleach, Dark food coloring, Water, Glass, Eyedropper, Toothpaste, Egg, Vinegar, Glass, Water, Toothbrush, Spoon, Colored nail polish, Plastic wrap, Periodic table, Scissors, Glue, Art supplies (markers, colored pencils, crayons)	
	Day 2	The Tooth, the Whole Tooth, . . . Element Book Group 17		
17	Day 1	He Never Argues - Group 18 (FMN) The Incredible Shrinking Balloon	Helium-filled balloons, Cloth tape measure, Freezer, Timer, Periodic table, Scissors, Glue, Art supplies (markers, colored pencils, crayons)	
	Day 2	Element Book Group 18		
18	Day 1	Putting It All Together (FMN) Make a Molecule Puzzle	Scissors, Glue, Card stock, Gumdrops, Toothpicks, Colored pencils, Large work surface, Sugar, Salt, Flour, Pepper, Peanut butter, Vegetable oil, Clear glass of water, Lettuce, Tomato, Carrot, Salad bowl	
	Day 2	Friendly Gumdrops Mixture or Compound?		
19	Day 1	Molecular Formulas Are Useful (FMN) Molecular Formulas Worksheet	Celery, Glass, Blue food coloring, Water	
	Day 2	The Celery Blues		
20	Day 1	Drawing Lessons (FMN) Drawing Lessons Worksheet	My Periodic Table, Blank paper, Chalkboard or dry erase board, White paper towel, Water, Shallow dishes, Tablespoon, Food color, Vegetable oil, Q-tips, Scissors, Ruler, Tape, Protected work surface	
	Day 2	Capillary Action in Action		
21	Day 1	Matter: An Introduction	My Periodic Table, Colored pencils, Blank paper, LEGO blocks	
	Day 2	Matter: An Introduction (continued)		
22	Day 1	The States of Matter (FMN)	Ice, Water, Pot, Heat source, Glass containers, Sealable baggies, Rock, Kitchen scale, Drinking straw (optional)	
	Day 2	Presto-Change-O Water		
23	Day 1	Let's Get to the Point (FMN) What Is the Point?	Crushed ice, Clear glass, Distilled water, Stove, Pot, Science thermometers, Pot Holder, Science encyclopedia or Internet access (optional), Jello, Cold and hot water, Peanut butter, Mayonnaise, Cups, Spoons, Bowl, Refrigerator, Plate	
	Day 2	State of Confusion		
24	Day 1	Solids Are Dense (FMN) Some Are Denser Than Others	Colored pencils, Medium-sized box or container, Stuffed animals, Bathtub or sink filled with water, Orange, Small plastic container, Marbles, Assortment of solids	
	Day 2	The Sinking Tub Boat		

Week	Day	Lesson / Lab	Supplies Needed For the Week	Dates / Notes
25	Day 1	Molecules Stick Together (FMN)	Sugar, Water, Saucepan, Stove top, Glass jar, Rough string or yarn, Wooden spoon, Metal washer or a Lifesaver candy, Funnel (optional)	
	Day 2	A Big Rock Candy Mountain!		
26	Day 1	What Makes a Liquid a Liquid? (FMN) Liquids Are Dense Too	Corn syrup, Vegetable oil, Water, Clear container, Measuring cup, Food coloring, Chalkboard or sheet of paper, Slime (made in former lab)	
	Day 2	Drawing the States of Matter		
27	Day 1	Friendship of Oxygen & Hydrogen (FMN) Smart Ice	Freezer, Water, Empty plastic soda bottles, Blue food coloring, Blue crayon, Pan, Water, Clear glass tumbler, Insulated copper wire, 9-volt battery, Salt, Spoon	
	Day 2	The Breakup		
28	Day 1	What Makes a Gas a Gas? (FMN) Bubbles	Unopened can of soda, Clear glass, Popping corn, Pan to pop popcorn, Knife, Oil, Heat source for popping corn, Kite, Windy day	
	Day 2	Popping Corn Let's Go Fly a Kite!		
29	Day 1	The Air You Breathe (FMN) Air Takes Up Space	Balloon, Empty 2-liter soda bottle, Glass jar with a lid, Tub or sink full of water	
	Day 2	Why Do Boats Float?		
30	Day 1	Changes (FMN) Physical or Chemical?	Piece of paper, Alka-Seltzer tablet, Water, Match or lighter, Kool-Aid, Sugar, Glass, Container to freeze Kool-Aid, Pitcher, Bubble blowing solution and wand, Cast-iron skillet or other fireproof container, Scissors	
	Day 2	Detecting Changes		
31	Day 1	Chemical Reactions Let's Heat Things Up	LEGO® blocks, Chalkboard or dry erase board, Science or kitchen thermometer, Yeast, Hydrogen peroxide, Clear container, Baking soda, Lemon juice, Spoon	
	Day 2	Let's Cool Things Down		
32	Day 1	Some Like It Sour, Some Don't (FMN) Step 1	Red cabbage, Knife, Distilled water, Strainer, Glass quart jar with a lid, Large container, White coffee filters, Cookie sheet, Bowl, Rubber gloves, Baggie, Vinegar, Ammonia, Lemon juice, Baking soda, Laundry detergent, 7-Up, Salt, Clear plastic disposable cups, Protected work surface	
	Day 2	Step 2		
33	Day 1	Hydrogen & Oxygen & Hydrogen. . . (FMN) Let's Make Water	White vinegar, Baking soda, Cabbage indicator, Indicator paper, Scissors, Small cups, Paintbrush, Water, Protected surface	
	Day 2	Painting Magic		
34	Day 1	pHunny pHriends (FMN) Make a pH Scale	Red, blue, and white watercolor or acrylic paint; Paintbrush; Paint pallet; Water; Cabbage indicator paper; Scissors; Knife; Lemon; Grapefruit; Lime; Orange; Cherry tomato; V-8 juice; Cup; Glue or tape	
	Day 2	pHun with Acids		
35	Day 1	Building Teeth (FMN) Precipitates	Baggie, Chalk, White vinegar, Container with a spout, Plastic wrap, Clear glass container, Bowl, Warm water, Baking soda	
	Day 2	Precipitates (continued)		
36	Day 1	Combustion Action (FMN) Playing with Fire	Glass jars, Roasting pan, Sand, Votive candles, Matches or a lighter, Stopwatch, Dollar bill, Tongs, Salt, Rubbing alcohol, Water, Measuring cup, Sink or other nonflammable surface	
	Day 2	Burning Money		

Further Reading and Exploring

The Usborne Science Encyclopedia is a good general science reference.

Unit 1 – WHAT IS CHEMISTRY?

How to Think Like a Scientist by Stephen P. Kramer
I Can Be a Chemist by Paul Sipiere
What Is a Scientist? by Barbara Lehn

Unit 2 – STARTING SMALL

Atoms (Simply Science) by Melissa Stewart
What Are Atoms? (Rookie Read-About Science) by Lisa Trumbauer
What's Smaller Than a Pygmy Shrew? by Robert E. Wells
Can You Count to a Googol? by Robert E. Wells

Unit 3 – THE CHEMIST'S ALPHABET DEFINED

The Periodic Table (True Books) by Salvatore Tocci
What Is Mass? By Don L. Curry
Grab a Seat at the Periodic Table!: A Chemical Mystery by Laura Layton Strom
The Periodic Table of Elements (Reading Essentials in Science) by Jenny Karpelenia
Sorting the Elements: The Periodic Table at Work (Let's Explore Science) by Andrew Solway

Unit 4 – THE CHEMIST'S ALPHABET APPLIED

Hydrogen and the Noble Gases by Salvatore Tocci
Calcium and the Alkaline Earth Metals by Nigel Saunders
Neon and the Noble Gases by Nigel Saunders
Carbon and Group 14 Elements by Nigel Saunders
Sodium and the Alkali Metals by Nigel Saunders
Oxygen and the Group 16 Elements by Nigel Saunders
Carbon by Salvatore Tocci
C Is for Carbon by Marilee Summers
Nitrogen by Salvatore Tocci
Oxygen by Salvatore Tocci
Sodium by Salvatore Tocci
Aluminum by Salvatore Tocci
Silicon by Salvatore Tocci
Chlorine by Salvatore Tocci
Nitrogen and Group 15 Elements by Nigel Saunders
Fluorine and the Halogens by Nigel Saunders

Unit 5 – MOLECULES RULE

Atoms, Molecules, and Quarks by Melvin Berger

Unit 6 – WHAT'S THE MATTER?

What Is the World Made Of? All About Solids, Liquids, and Gases by Kathleen Weidner Zoehfeld and Paul Meisel
What Is Matter? by Don L. Curry
What Is Volume? by Lisa Trumbauer
What Is Density? by Joanne Barkan
Everything Is Matter! by David Bauer and David Lewis
Matter: Solids, Liquids, and Gases by Mir Tamim Ansary
Will It Float or Sink? by Melissa Stewart
Matter: See It, Touch It, Taste It, Smell It by Darlene R. Stille and Sheree Boyd

What's the Matter in Mr. Whiskers' Room? by Michael Elsohn Ross and Paul Meisel
Solids, Liquids, and Gases by Ginger Garrett
Change It!: Solids, Liquids, Gases and You by Adrienne Mason and Claudia Dávila
Freezing and Melting by Robin Nelson
Air Is All Around You by Franklyn Mansfield Branley and Holly Keller
A Drop Of Water by Walter Wick
Ask Magazine, April 2007, "The Wonder of Water"
KIDS Discover Magazine, May 2007, "Water"
I Get Wet by Vicki Cobb

Unit 7 – REACTIONS IN ACTION
Acids and Bases by Carol Baldwin

Web Suggestions

Visit Pandia Weblinks for links to suggested sites: www.pandiapress.com/weblinks-chem-1

Unit 1

What Is Chemistry?

I Am a Chemist

It started with an atom
To this I must confess.
Then it was a molecule
With not one atom less.

Some kids they build with Legos
And some with Playmobile.
But me
Oh, I like atoms
I build things that are real.

I may be only eight
But I already know my fate.
It has been addressed
Yes, I am a chemist.

Name ______________________________ Date __________________

For my notebook

What Is Chemistry?

Have you ever heard grown-ups talking about chemicals in the food you eat, the water you drink, or in the air you breathe? Can I tell you a secret? Everything around you that you can see, and even the things you cannot see, are chemicals. Everything in food is a chemical. Water, itself, is a chemical. The air we breathe is made up of chemicals. If your sister has blue eyes and you have brown eyes, it is because of chemicals. This book is made of chemicals. When grown-ups talk about chemicals, they are usually talking about chemicals that are bad for you. Most chemicals, though, are either good for you or won't hurt you at all. Okay, maybe if you drop this book on your foot it would hurt, but you get my point.

Chemistry is the science that studies chemicals. Since everything is made of chemicals, the science of chemistry interrelates with all other types of science. When you study why a plant grows or how your five senses work in life science, you are also studying chemistry. When you study volcanoes erupting in earth science or how the planets are different from one another, you are studying chemistry. Why a rainbow forms and makes colors, or why some things float and some things sink, is physics and chemistry, too. Medical science is as much chemistry as it is biology. You have probably been studying chemistry for a long time and didn't know it!

A chemist is a person who studies how chemicals interact. Chemists ask questions like: Why do lemons taste sour and smell lemony? How do plants die in the winter and come back in the spring? How can a group of chemicals come together and make a cat? Why do volcanoes erupt? What are stars made of? Why do some people have blue eyes? Why are blue and green next to each other in a rainbow? What kinds of medicine will best help kids when they catch colds? There are so many questions. This year you are going to ask and answer a lot of questions, because this year you are going to be a chemist.

What Is Chemistry? Lab #1: Be a Chemical Detective -instructions

Materials:
- Lab sheet, pencil
- Box of cereal that has several ingredients

Aloud: Today you will be a chemical detective. You will be looking for chemicals in common, everyday things. That shouldn't be very hard since everything around you is made of chemicals.

Procedure:
1. In a home situation: Have students go through the house with the lab sheet and a pencil. They will look at a box of cereal and read the list of ingredients. Then they will search items in the rest of the house, looking for chemicals listed on labels. Good places to look include your laundry room, bathroom, garage, and pantry.
2. In a classroom situation: You will need to make sure you have cereal boxes and several other items with chemicals on hand.

Instructor's Notes:
- The list of ingredients on boxes of your food items are a good place to look for many of the chemicals.
- Cleaning fluids are a good place to look for dangerous chemicals. Bleach, ammonia, toilet bowl cleaner, and lighter fluid are dangerous.

Possible Answers:
Cereal box:

I have heard of these - rice, granola bar pieces, sugar, salt, oats

I have never heard of - red dye #50, annatto, apple puree concentrate

Around your house or school:

Four chemicals you have never heard of - smoke flavor, sodium benzoate, sodium pyrophosphate, sodium stearoyl lactylate

Four things made of chemicals I use every day - air, water, toilet paper, your brain

Two chemicals that are dangerous - bleach and lighter fluid

A chemical that makes your car go - gasoline

Your favorite chemicals to drink - apple juice

Your favorite chemicals to eat - coconut

Two things that are bigger than a microwave made from lots of chemicals - refrigerator, television, my mom (more than two, but all are good answers)

Two things smaller than a toaster made from lots of chemicals - knife and a mouse

If you were a chemist, what question would you try to answer? - "What are black holes made of?"

Name ______________________________ Date __________________

What Is Chemistry? Lab #1: Be a Chemical Detective

Chemicals in a Box of Cereal

I have heard of these:	I have never heard of these:

Around Your House or Classroom

Four chemicals I have never heard of:

1.	2.
3.	4.

Four things made of chemicals I use every day:

1.	2.
3.	4.

List two chemicals that are dangerous:

1.	2.

A chemical that makes your car go =

Your favorite chemicals to drink (hint: mine is water) =

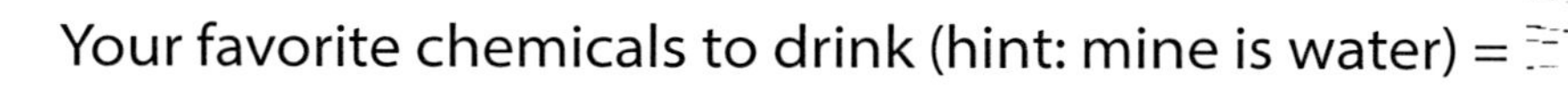

Your favorite chemicals to eat (hint: mine is chocolate) =

Two things bigger than a microwave that are made from lots of chemicals:

1.	2.

Two things smaller than a toaster that are made from lots of chemicals:

1.	2.

Bonus: If you were a chemist, what question would you try to answer?

What Is Chemistry? Lab #2: Telling Things Apart - instructions

Materials:

- Lab sheets (two pages), pencil
- ½ cup Confectionery sugar
- ½ cup Baking powder
- ½ cup Baby powder*
- ¾ cup Water
- ¾ cup Vinegar
- ¾ cup Cooking oil—any type
- Blue food coloring
- Three teaspoons
- Container to pour liquids from
- Three clear glasses
- Three dishes to hold confectionery sugar, baking powder, and baby powder
- Kitchen towel, sink, and dish soap to clean glasses as you go along

* If you have concerns about using baby powder, cornstarch may be substituted, but the results/observations will be different.

Aloud: Sometimes things look the same even though they are different. That can make it hard to tell them apart. When chemists want to know if one thing is different from another, they use two main types of tests. The first type of testing is physical. In physical tests, observations are made about the physical properties of things. Physical properties are how things look, smell, taste, and feel. A chemist would not taste things if she did not know what they were, though, because she would not know if they were dangerous.

The second type of testing is chemical. With chemical tests, a chemist performs experiments on things to see how they behave chemically. She will look at things to see what happens when she puts them in water, in oil, or in an acid, such as vinegar.

In this experiment, you will learn how you can tell different things apart using physical tests and chemical tests. You will take three similar-looking powders and examine them. You will run a series of physical and chemical tests on baby powder, confectionery sugar, and baking powder. Your goal is to observe the physical differences between these three powders and to observe the different ways that these three things behave when you perform chemical tests on them. You begin this lab, as you do most of the labs in this book, by writing a hypothesis for the lab. A hypothesis is your best guess about the outcome of the lab based on what you have learned so far. The plural form of the word hypothesis is hypotheses.

Procedure:

1. Complete the hypothesis portion of the lab report.
2. Get out all the materials. The liquids do not need to be measured out at this time.

Physical Tests:

3. Measure about ½ cup each of confectionery sugar, baking powder, and baby powder into separate dishes.
4. Have students look at each of the powders and note the differences they observe using the physical tests of sight, touch, and smell on the lab sheet. When smelling these, be careful not to breathe in or out too hard.

Chemical Tests:

5. Use a different spoon for each powder. Measure a leveled spoon of each powder into a separate glass.
6. Add about $^{1}/_{8}$ cup of water to each powder. Stir them with a clean spoon and record observations on the lab sheet.
7. Clean out the glasses and dry them.
8. Measure a leveled spoon of each powder into separate glasses.
9. Add about $^{1}/_{8}$ cup of vinegar to each. Stir them with a clean spoon and record observations on the lab sheet.
10. Clean out the glasses and dry them.
11. Measure a leveled spoon of each powder into separate glasses.

(continued on the back)

12. Add 3 drops of food coloring to each glass.
13. Add about 1/8 cup of water to each. Stir them with a clean spoon and record observations on the lab sheet.
14. Add about 1/8 cup of vinegar to each of the glasses. The baking powder mixture might fizz out of the glass.
15. Do they behave differently than they did with only one of the liquids? Record observations on the lab sheet.
16. Clean out the glasses and dry them.
17. Measure a leveled spoon of each powder into separate glasses.
18. Add about 1/8 cup of oil to each. Stir them with a clean spoon and record observations on the lab sheet.
19. Add about 1/8 cup of vinegar to each glass. Wait a minute or so for this part to see what happens. Do not stir. Record observations on the lab sheet.
20. Add 3 drops of food coloring to each glass. Stir them with a clean spoon and record observations on the lab sheet.
21. Raise the glasses to look at the bottoms.

Instructor's Notes:

- I have given specific measurements for the amounts of ingredients, but that is only to make sure you have enough of everything. These measurements do not need to be very accurate.
- Have students fill out their lab sheets as they go along in the experiment. A lot happens in this experiment.
- Use a clean spoon for each of the powders during the Chemical Tests part of this experiment.
- When you are doing this experiment, it is advised that you pick the glasses up to see what is happening on the bottom of each glass.
- The answers for this experiment are only possible answers. Your students will be able to tell the three powders apart, but their descriptions of what is happening might be different from mine.
- I have intentionally not gone into the chemistry involved in this experiment. That would confuse the intention of this experiment at this level. My intention was to teach the difference between physical and chemical tests and to demonstrate how a chemist might go about telling different things apart. This experiment emphasizes observation skills and attention to details. These are very important skills for studying science.
- 1/8 cup = 1 ½ tablespoons

Possible Answers/Observations:

Physical Tests

Sight			Smell			Touch		
baby p.	sugar	baking p.	baby p.	sugar	baking p.	baby p.	sugar	baking p.
white	*white*	*white*	*good*	*none*	*none*	*soft*	*soft*	*grainy*

Chemical Tests

baby p. + water	sugar + water	baking p. + water
does not dissolve	*dissolves*	*bubbles/fizzes*
baby p. + vinegar	sugar + vinegar	baking p. + vinegar
does not dissolve	*dissolves*	*fizzes a lot*
baby p. + food color + water	sugar + food color + water	baking p. + food color + water
scummy and lumpy	*blue/ mixes with food color*	*bubbly blue*
+ vinegar	+ vinegar	+ vinegar
does not dissolve	*dissolves and clear blue*	*bubbly blue*
baby p. + oil	sugar + oil	baking p. + oil
does not dissolve	*does not dissolve*	*does not dissolve*
+ vinegar	+ vinegar	+ vinegar
oil on top, baby p. does not dissolve	*oil on top, sugar dissolves*	*bubbly and oily*
+ food color	+ food color	+ food color
food color dots in oil	*oil on top of blue liquid*	*blue, bubbly, and oily*

NAME ______________________________ DATE ________________

What Is Chemistry? Lab #2: Telling Things Apart - page 1

Hypotheses: Write your best guess to the following:

1. When I look at the physical properties of confectionery sugar, baby powder, and baking powder, I think I ________________ be able to tell them apart. (will/will not)

2. I think that confectionery sugar, baby powder, and baking powder will be ______________ to tell apart using chemical tests. (easy/hard)

Results/Observations:

Physical Tests

Sight			Smell			Touch		
baby powder	sugar	baking powder	baby powder	sugar	baking powder	baby powder	sugar	baking powder

Chemical Tests

baby powder + water	confectionery sugar + water	baking powder + water
baby powder + vinegar	confectionery sugar + vinegar	baking powder + vinegar
baby powder + food color + water	confectionery sugar + food color + water	baking powder + food color + water
+ vinegar	+ vinegar	+ vinegar
baby powder + oil	confectionery sugar + oil	baking powder + oil
+ vinegar	+ vinegar	+ vinegar
+ food color	+ food color	+ food color

What Is Chemistry? Lab #2: Telling Things Apart - page 2

Discussion and Conclusion:
Was there something special about each powder that helped you to tell them apart? Circle all the answers that fit for each powder. There are extra spaces for you to write your own observations.

1. Baby powder

is white | smells good | is soft

is grainy | fizzes | mixes with food color

dissolved in water | dissolved in vinegar | dissolved in oil

does not dissolve in water | does not dissolve in food color

does not dissolve in oil ________________ ________________

2. Confectionery sugar

is white | smells good | is soft

is grainy | fizzes | mixes with food color

dissolved in water | dissolved in vinegar | dissolved in oil

does not dissolve in water | does not dissolve in food color

does not dissolve in oil ______________ ______________

3. Baking powder

is white | smells good | is soft

is grainy | fizzes | mixes with food color

dissolved in water | dissolved in vinegar | dissolved in oil

does not dissolve in water | does not dissolve in food color

does not dissolve in oil ________________ ________________

What Is Chemistry? - Crossword Vocabulary Review

Across

5. An educated guess about the results of an experiment.
6. The science that studies chemicals.

Down

1. A chemist performs this to test how things behave chemically. (Two words)
2. A person who studies how chemicals interact.
3. A chemist performs this to test the physical properties of something. (Two words)
4. Chemists study these.

Unit 2

Starting Small

The Atom Song

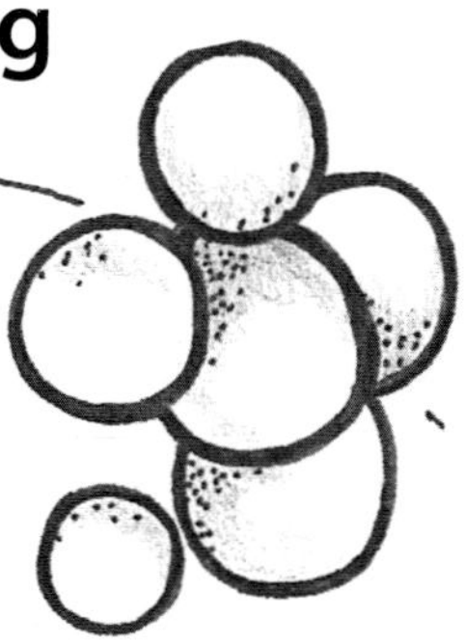

An atom
is so small
you can't see it,
not at all!

But without atoms
none of us
would be.
They make up
everything
you see.

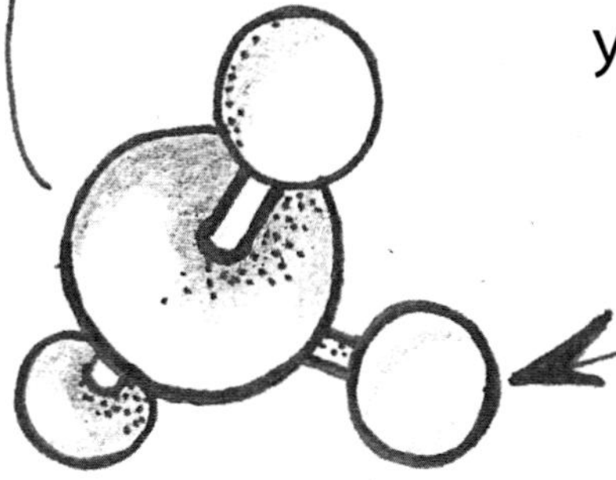

NAME ______________________ DATE ______________

For my notebook

The Atom

Have you heard of atoms? Did you know that everything in the world and the universe is made of atoms? Atoms are the basic building blocks of everything you see, including yourself. That means even cells are made of atoms. You remember what cells are, don't you? They are the building blocks of living things, and atoms are the building blocks of them. Atoms are like the Legos of the universe, only atoms are a lot smaller than Legos. They are so small that a person who weighs 75 pounds would have about 3,500,000, 000,000,000,000,000,000,000 (three octillion, five hundred septillion) atoms in his body! Try writing that number down; it's 35 followed by twenty-six zeros.

Thousands of years ago, the ancient Greeks thought a lot about how things are made. About 2,400 years ago, a Greek named Democritus (dih-MOCK-rih-tuss) said that everything was made from particles, called atoms. He thought that all things could be broken down into smaller and smaller pieces until you got to atoms. Democritus also thought atoms moved all the time and that they could join with each other.

The problem with Democritus's theory about atoms was that at that time, there was no scientific way to prove that atoms exist. Atoms are so small that people cannot see them without using a special type of microscope called a scanning-tunneling microscope. There were no scanning-tunneling microscopes 2,400 years ago. Most people living then found it hard to believe in something they could not see. That meant most of the people alive when Democritus was alive did not believe in atoms.

Today we know that Democritus was right. All things are made of atoms. He was right that atoms move all the time. He was also correct that atoms join together. When atoms join, they make molecules.

Move your hands in the air. As you move your hands through the air, you are hitting atoms and molecules. You cannot see them, but they are there. Air is mostly made of two types of atoms whose names are nitrogen and oxygen. Water is made of atoms, too. Water is made of two types of atoms called hydrogen and oxygen. Everything is made of atoms!

The Atom Lab #1: Are Atoms Small? - instructions

Materials:

- Lab sheet, pencil
- Five balloons that have not been inflated
- Almond extract
- Water
- Cinnamon
- Lemon, orange, or peppermint extract
- Magnifying glass
- ½ teaspoon measuring spoon
- Permanent marker

Part 1:

Aloud: Atoms are really small. Think of the smallest thing you have ever seen with your own two eyes. Atoms are a lot smaller than even that. Look at your lab sheet. Do you see the dash under the magnifying glass? How many atoms do you think are in that dash?

Procedure:

Have students trace over the dash on the lab sheet with a pencil, and examine it with a magnifying glass. Wait for students to write a guess about the number of atoms.

Aloud: There are 40,000,000 (40 million) atoms in that dash! Atoms are small, but everything is made of them. The next time you go outside, look at all the different things in the world that are made of atoms. If it is a sunny day, remember even the sun is made of atoms. If it is rainy or cloudy, remember that the clouds and the raindrops are all made of atoms. Oh, by the way, a raindrop has about 5,000,000,000,000,000,000,000 (5 sextillion) atoms in it. If you catch one on your tongue, think about that! Do you remember what kind of atoms are in raindrops? There are hydrogen atoms and oxygen atoms in raindrops because raindrops are made of water.

Instructor's Notes:

The dash is 2mm long. There are about 20 million carbon (graphite) atoms in a pencil dash that is 1 mm.

Part 2:

Aloud: What does the outside of a balloon smell like? Would you say sort of rubbery or like nothing at all? What if you put something with a strong scent or smell into a balloon? Would you be able to smell what's in the balloon if you inflated it? How could you? Maybe you could put a small hole in it. The problem with that is, if a balloon had a hole, it wouldn't hold air, would it?

Today, you are going to smell five balloons. Each balloon has something different in it. You will see if you can smell the scent atoms through the balloons.

Balloons are made of atoms like everything else in the world. The things you will be putting into the balloons are made of atoms, too. Do you think the scent atoms will be small enough to go through the atoms of the balloon?

Procedure (read over the entire procedure before starting the lab):

1. Complete the hypothesis portion of the lab sheet.
2. Before inflating the balloons, have students examine them. They should smell them and check them for holes with a magnifying glass. If they find a hole in a balloon, discard it and get another with no holes. Blow up one of the balloons and have the students examine this balloon with the magnifying glass. They are checking it for holes.

(continued on the back)

3. Do this next step before inflating the rest of the balloons and out of sight of your students. Pour water, cinnamon, almond extract, and the other type of extract in four different balloons. After each addition, blow up the balloon and tie off tightly. Do not over-inflate the balloons. If you do, they could pop and you will have a mess. Be careful not to get anything on the outside of the balloons or your hands. If you do get something on the balloon, wash it off with soap and let it dry. Label the balloons 1, 2, 3, and 4, or you can use different-colored balloons for identification. The students will guess what is in them. The rest of the experiment is done in front of the students. Shake each balloon for 30 seconds starting with the balloon that has only air in it. Have students smell the outside of the balloon. Have them record results on the lab sheet.

Instructor's Notes:

- If you use peppermint extract, put it at the end of the experiment. It smells so strong that it can affect how the unscented balloons smell. You might want to leave it in another room until all the other balloons have been tested.
- Cinnamon and vanilla extract can be seen through light-colored balloons. Try using a dark-colored balloon for these scents.

Possible Answers:

Results / Observations:
Before being inflated, the balloons should smell like nothing, or rubbery, or like a balloon.
Before and after the balloons are inflated, students should not see any holes in the balloons.

Data Table:
Students should fill in the part of the data table where they guess what the balloons have in them. You should help them fill in the part that tells what was really in the balloons.

They should smell both extracts and cinnamon. They should not smell anything from the balloon with air in it and the balloon with water in it. They might correctly guess the balloon with water because they will be able to hear that it has liquid in it.

Discussion/Conclusion:
They should have smelled all three things that had a scent. From this, students should have learned that the scent molecules and atoms are small enough to travel between the atoms that make up the balloon.

NAME ______________________________ DATE ________________

The Atom Lab #1: Are Atoms Small?

Part 1:

–

I think there are ______________________________ atoms in that dash.

Part 2:

Hypotheses (circle your answers):

I think scent molecules are small enough to travel through the molecules of the balloons and that I will smell the scents put in the balloons.

Yes No I don't know

I think the balloons with air and water will smell the same as they did before being inflated.

Yes No I don't know

Results / Observations:

Before being blown up, my balloons smelled like ________________

When I looked at the balloon with my magnifying glass, I saw:

Before being blown up ______________________________

After being blown up ______________________________

Data Table	The balloon smelled like	What was in the balloon?
balloon filled with air		*air*
balloon #1		
balloon #2		
balloon #3		
balloon #4		

Discussion and Conclusion:

Did you smell any of the things put into the balloons?

Which ones did you smell?

What did the scents teach you about the size of atoms and molecules?

The Atom Lab #2: Do Atoms Move? - instructions

CAUTION: This lab involves handling very hot water.

Materials:

- Lab sheet, pencil
- Color pencil or crayon (same color as the food coloring)
- 3 cups of Water at three different temperatures:
  1) Chilled (Put ice and water into a container and drain off the water for use.)
  2) Room temperature
  3) Very hot (just been boiled)
- Food coloring (Use the same color and amount for each test. A darker color is better.)
- Three clear glasses, the same size
- Thermometer, science or kitchen-type
- Stopwatch or a timer that counts in seconds

Aloud: When you look at a drop of water, can you tell that the hydrogen and oxygen atoms in it are moving? Well, they are moving, and very fast, too. In this lab, you are going to drop food coloring into water. You will not see a single food color atom move through the water; atoms are too small to see by themselves. But you can see a group of food color atoms move through the water. When you put the drops in the water, the food coloring will mix with the water without you stirring it. When things mix without being stirred, it is called diffusion. Temperature can affect how fast atoms and molecules mix with each other. The water in each glass will be a different temperature. Do you think the molecules will diffuse faster in the hot water or the cold water?

Procedure:

1. Complete the hypothesis portion of the lab sheet.
2. Measure one cup of each temperature of water into three clear glasses.
3. Right away, measure the temperature of each glass of water. Do this very carefully so you don't stir the water. (To prevent the thermometer from shattering, allow it to cool for a few seconds between the hot and cold water.) When the thermometer stops moving up or down, record the temperatures on the lab sheet.
4. Carefully drop 5 drops of food coloring into each glass of water.
5. Immediately observe what happens in each glass and record observations on the lab sheet. Observations should be recorded in words and pictures.
6. Wait 2 minutes. Measure the three temperatures again. Record observations on the lab sheet.
7. Wait 30 minutes. Measure the three temperatures again. Have there been any changes?
8. Complete the lab report.

Aloud: When you can see things diffuse, you are watching molecules and atoms in motion. Heat can make atoms and molecules move faster. You used colored molecules in this experiment so it would be easy to see them move through the colorless water. But atoms are moving all the time, even when you can't see them.

Instructor's Notes:

- Make sure each glass has the same amount of water. If you use more or less than a cup of water, the rate of diffusion will be affected.
- Make sure the water is not stirred, or otherwise moving, when you carefully drop in the food coloring. You want the atoms to mix through diffusion, not from stirring.
- When food coloring is put in the hot water, it diffuses very quickly. Make sure students are watching the experiment right from the start.
- Thirty minutes might not be enough time for the food color to diffuse completely through the water in the room temperature and cold water. Try leaving the glasses sitting out until the color diffuses completely.

(continued on the back)

Possible Answers:

Hypotheses:
The correct answers are yes, yes, hot.

Results:
Data Table
The temperatures will vary.

Observations:
Each square represents a glass of water with food coloring in it. The coloring in each square should look similar to the diffusion pattern in each glass of water + food coloring at the specified time.

Conclusion:
The atoms diffused fastest in hot water.
The atoms diffused slowest in cold water.

NAME ________________________________ DATE ________________

The Atom Lab #2: Do Atoms Move?

Hypotheses:

Do you think you will see the food color atoms diffuse (move) through the water?

Yes   No   I don't know

Do you think the temperature of the water affects the rate of diffusion (how fast things move) in the water?

Yes   No   I don't know

I think atoms move faster when they are ________________.

cold   room temperature   hot

Results:

Temperature:	Chilled water	Room temperature water	Hot water
Start			
2 minutes			
30 minutes			

Observations: Color the squares to show what is happening to the food coloring in each glass of water.

Chilled water			Room temperature water			Hot water		
Start	2 min	30 min	Start	2 min	30 min	Start	2 min	30 min

Conclusion: Circle the correct word(s) to complete each sentence.

The atoms diffused fastest in   cold   room temperature   hot   water.

The atoms diffused slowest in   cold   room temperature   hot   water.

What's in an Atom?

So what's in an atom?
Let's start
And learn
Each part.

Proton,
Electron,
Neutron,
Yeah.

Inside the nucleus,
There are two kinds of things,
The neutral neutron
And the positive proton.
They're good friends,
Like you and me.
They don't like to be seen separately.

Proton,
Electron,
Neutron,
Yeah!

Then there's a little guy
Orbiting around.
He's really fast
But he doesn't make a sound.
He's the electron
And he's negatively charged.
He's really small.
He's not at all large.

Proton,
Electron,
Neutron,
Yeah!!

Proton,
Electron,
Neutron,
Yeah!!!

Proton,
Electron,
Neutron,
Yeah!!!!

Parts! Poster -instructions

Materials:
- "Parts of an Atom" poster (p. 47), 1 per student
- Crayons or colored pencils - purple, red, blue, green, and orange
- Construction Paper - one 8 ½ " x 11" piece
- Glue
- "What's in an Atom?" poem (p. 44)

Hand out the "Parts of the Atom" poster found on page 47. Students should follow along on it while you read below.

Aloud: The picture on the poster shows the parts of an atom. Atoms are very small, but there is something even smaller than atoms. Atoms are made of three main parts, and the parts that make up atoms are smaller than atoms. These three parts are called <u>protons</u> (proh-tonz), <u>neutrons</u> (noo-tronz), and <u>electrons</u> (ee-lek-tronz).

Let's learn what an atom looks like. I am going to read a description of the parts of an atom to you. I want you to follow along on your Parts of an Atom poster using your crayons.

Look in the center of the atom. Do you see the four circles in the center of the atom? This center part of an atom has a special name. It is called the <u>nucleus</u> (noo-klee-uhss). With a purple crayon, draw one tight circle around all four circles in the nucleus. Can you find the word "nucleus" in the word box? Shade the word "nucleus" purple. The nucleus is made of things called protons and neutrons.

Find the two neutrons by looking for the letter "n" inside two of the circles in the nucleus. Color the neutrons blue. Look at the neutrons. They look the same as each other. That is because all neutrons are the same as each other. Find the word "neutron" in the word box. Shade the word "neutron" blue.

The two protons in the nucleus have a "p+" inside their circles. Color the protons red. Look at the protons. They look the same as each other. That is because all protons are the same as each other. Find the word "proton" in the word box. Shade the word "proton" red.

An electron is a small particle that orbits around the nucleus. There are two of them; they have an "e-" in their circles. Can you find both of them? Color them green. Look at the electrons. They look the same as each other. That is because all electrons are the same as each other. Shade the word "electron" green.

The <u>energy level</u> is where you find the electrons of an atom. Trace over the energy level, the big circle the electrons are in, with an orange crayon. Shade the words "energy level" at the top of the page orange.

Now there is one last part of your atom I want you to notice. Well, it's actually not a part, but rather the lack of a part. Let me explain. Do you notice what is between the nucleus where the protons and neutrons are and the energy level where the electrons are? Nothing! Atoms have a whole lot of empty space in them. In fact, empty space is the biggest "part" of an atom.

Cut out your Parts of an Atom poster and glue it onto a piece of construction paper. Hang it on your wall to help you remember the parts of the atom while you study chemistry this year.

The atom you just colored is a <u>helium</u> (HEE-lee-em) atom. It is a special type of atom. Helium is used in balloons to make them float. It is also what they put in blimps to make them lighter than air.

Instructor's Notes:
- The Parts of the Atom poster created today is used as a reference by students throughout this book.
- Recite the "What's in an Atom?" poem to help students remember the parts of an atom.
- Throughout this course I will be referring to electrons being on energy levels as they circle the nucleus in an atom. Traditionally these circles were called "orbits," but recent atomic discoveries have led to the more accurate term "energy level." Whether energy level or orbit, when teaching about atoms we draw models with electrons neatly circling the nucleus. But the fact is, the placement of electrons is more accurately described as a cloud, and the location of electrons in a cloud is determined by a probability function. For this age-group, however, this concept is best taught as tidy energy levels.

(continued on the back)

For More Lab Fun:

If you have a group of three or more students, try acting out an atom while reciting the "What's in an Atom?" poem. Have students play the parts of the "proton" and the "neutron" standing very close together in the center, and the "electron" orbiting around them in the "energy level." Clap and shout the chorus.

Recite "What's in an Atom?" at the start of each science class for the next few weeks until students know the parts of an atom well. Then review the poem periodically throughout the school year.

NAME ______________________________ DATE ____________________

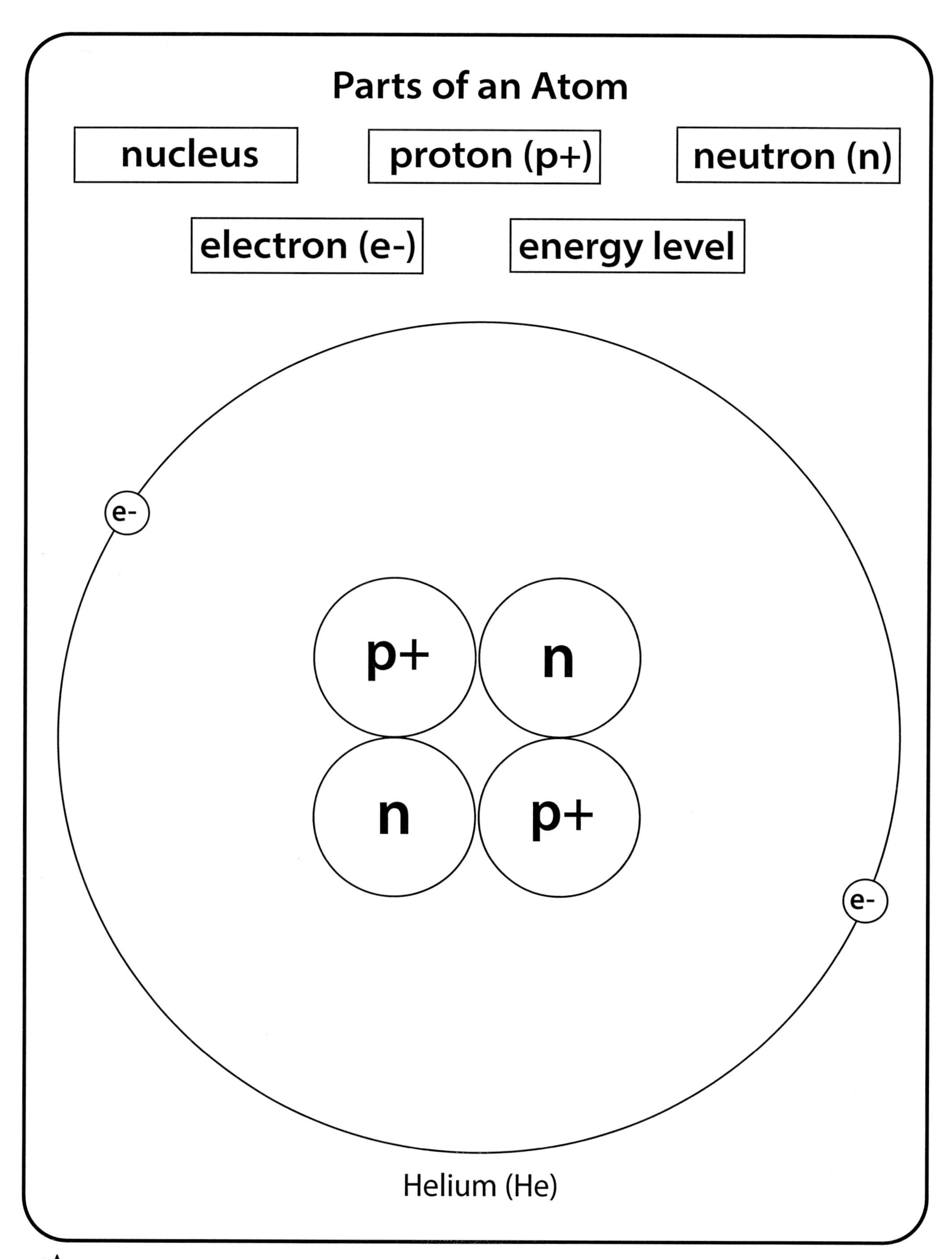
Parts of an Atom
nucleus
proton (p+)
neutron (n)
electron (e-)
energy level
e-
p+
n
n
p+
e-
Helium (He)

Parts! Lab: Let's Be Positive - instructions

Materials:
- Copy of lab sheet, pencil
- Inflated balloon
- Wall
- Mirror
- Carpet or sofa covered in cloth material (leather sofas won't work)
- Completed "Parts of an Atom" poster

Aloud: Have you ever gotten a shock when you touched something or someone? Have you taken something out of the dryer and had it cling to you? These things happen when enough electric charge builds up and moves from one thing to another.

Are you thinking, "Hey, wait a minute! Isn't charge what horses do at the start of a battle?" Maybe you are thinking, "Isn't charging what my mom does when she goes shopping?" You can see both of those types of charge. You can see the other type of charge, too.

The electrons orbit around the outside of the atom and sometimes you can rub them off onto something else. A charge can be either positive or negative. Do you notice the protons have "+" signs and the electrons have "-" signs in the atom on the "Parts of an Atom" poster? Protons have positive charge and electrons have negative charge. That is why there are "+" signs in the protons and "-" signs in the electrons.

When you rub a balloon on carpet or a sofa, electrons will rub off them and on the balloon. The balloon will have a negative charge from the electrons because of this. The positive part of your hair and the wall will be attracted to the negatively charged balloon. Opposites really do attract!

Procedure:
1. Before rubbing the balloon on anything, put the balloon next to a student's hair and the wall. Have students record their observations on the lab sheet.
2. Rub the balloon ONLY in one direction on your carpet or sofa. Rub it five to ten times. Be careful not to pop the balloon.
3. After rubbing the balloon, hold it next to a student's hair and let him look in a mirror. After that, touch the balloon to a wall. Have students record their observations on the lab sheet.

Aloud: When the balloon is rubbed on the sofa/carpet, electrons rub off the sofa/carpet and onto the balloon. The extra electrons on the balloon attract the protons in your hair and on the wall. Try it on more things. If you rub the balloon again, though, make sure you do not rub it in a different direction.

Instructor's Note:
- If it is humid or rainy where you live, this might not work. Wait for a drier day.

Possible Answers:
Before rubbing the balloon: nothing happened, nothing happened
The pictures should show hair standing up and a balloon sticking to the wall.

NAME ______________________________ DATE ____________________

Parts! Lab: Let's Be Positive

Before rubbing the balloon on the carpet or sofa, . . .

when I held the balloon close to my hair

____

____

____.

when I held the balloon close to the wall ____________

____

____.

Results: Draw pictures to show what happened after rubbing the balloon.

When I held the balloon close to my hair, it looked like this:

When I held the balloon close to the wall, it looked like this:

NAME ______________________________ DATE ________________

For my notebook

Types!

The atom on your "Parts of an Atom" poster is a special type of atom. It is a helium atom. There are over 100 different types of atoms. Isn't that a lot? How could that be if all atoms are made of the same three things—electrons, protons, and neutrons? It is true, though. Just look around you. Isn't it incredible that all the things you can see are made from the same three things?

So, if everything is made up of the same three things—protons, electrons, and neutrons, then what makes one type of atom different from another? Think of a piece of aluminum foil and a pencil lead. Aluminum foil is made of aluminum atoms, and a pencil lead is made of carbon atoms. Remember, an electron in an aluminum atom is exactly the same as an electron in a carbon atom. The protons are all the same as each other in both atoms and so are the neutrons. Aluminum and carbon look so different from each other, it is hard to believe they are made from the same things, isn't it? An atom of aluminum has more protons, neutrons, and electrons than an atom of carbon, and that is all that makes it different. Wow!

Aluminum foil is made of LOTS of aluminum atoms. The tip of the pencil is made of LOTS of carbon atoms. A group of the same type of atoms is called an element. When you look at aluminum foil, you are looking at the element aluminum because there is only one type of atom in the foil (aluminum atoms) and there is more than one of them. The pencil lead is the element carbon because it is made up of only carbon atoms.

Types! Lab: The First Ten - instructions

Materials:

- Lab sheet, pencil
- Small piece of aluminum foil
- Pencil lead or a sharpened pencil
- One bag of mini marshmallows—all the same color (e.g. white)
- Two bags of regular-size marshmallows, 2 different colors—besides traditional white, you should be able to find chocolate (brown) and/or strawberry (pink) marshmallows
- Ten blank sheets of paper (8 ½ x 11)
- Large table, counter, or floor space where you can spread out 10 marshmallow atoms (at least 10 feet wide)
- Kitchen scale
- "Parts of an Atom" poster (completed) and "What's in an Atom?" poem
- Scissors
- Atomic Energy Levels Diagram (page 59)
- Periodic table found on the inside back cover of this book

Aloud: Look at the piece of aluminum foil* and the pencil lead*. Remember that the only difference between them is that aluminum atoms have more electrons, protons, and neutrons than the carbon atoms making up the pencil lead. Today you will make ten different types of atoms using marshmallows, and the only difference between them will be the number of protons, electrons, and neutrons. Each different type of atom in the universe has its own name. There are more than 100 and the number keeps growing. That means there are over 100 names for the types of atoms. Today you will make the first 10 types of atoms.

* There are impurities (small amounts of other types of atoms) found in aluminum foil and pencil leads. For the sake of teaching elements, I ignored this fact.

Procedure:

1. Let students examine and compare the aluminum foil and pencil lead.
2. Cut out the name squares of the different types of atoms on the lab sheet. The pronunciation for them is given as each type of atom is introduced. There is a number by each name. The numbers are the atomic numbers and also indicate the order in which you will make the atoms. The name with the number 1 by it, for example, is the atom that is made first.
3. Before beginning to build the atoms, use the "Parts of an Atom" poster that students made last week and "What's in an Atom?" poem to refresh their memories about what an atom looks like.
4. Decide which color of the regular-size marshmallows will be protons and which will be neutrons. The mini marshmallows will be electrons.
5. Students should build each atom on a separate sheet of blank paper as you read the scripted directions found on the next page. Work your way through building the atoms one at a time, beginning with hydrogen and ending with neon, placing the marshmallow protons, neutrons, and electrons in their proper places for each atom. As students make the atoms, have them put the matching atom name label on the sheet and write in the missing number of protons, electrons, and neutrons on the label.

Instructor Notes:

- This lab is in two parts: building the marshmallow atoms and then placing the marshmallow electrons in energy levels. You might choose to do both parts consecutively, in one long lab. If you choose to split this into two days, you have to rebuild the marshmallow atoms for the second day.
- Look at the periodic table found on the inside back cover of this book to assist you with placement of the marshmallow atoms on your work surface. The placement is important because it mimics the periodic table, which students will learn about later in this course.

(continued on the back)

- The number of neutrons does not increase by a consistent amount. Therefore, you will provide the number of neutrons for the student every time. The number of protons and electrons increases by one, going from one type of atom to the next. You will begin writing the numbers of protons and neutrons for students until a point, noted in the text. After that, you will discuss the pattern and students will help determine the correct number of protons and neutrons to write down.
- Many of the names of the atom types (element names) will not be familiar to your students. It is not the purpose of this lab to teach what these elements are. That is done in another unit. The names are given here as a way to distinguish the different types of atoms.
- As students build atoms, the number of protons and neutrons will increase. In order to fit all the marshmallows on the paper, you may have to stack the "protons" and "neutrons" on top of each other.

Aloud Part 1 - Building Atoms:

neutron 0
proton 1
electron 1

1. **Let's start making atoms. First is hydrogen (HI-dreh-jen).**
   - **Put a blank piece of paper on the top left side of your work surface. Cut out all of the atom labels found on the lab sheet. Glue or tape the name "hydrogen" onto the top of the paper.**
   - **You need one proton for your hydrogen atom. Where should you put it? (nucleus)**
   - **Now take one electron and put this orbiting the nucleus.**
   - **You have now made a hydrogen atom. Wait! Did I forget something? Where is the neutron? Guess what? Hydrogen does not have one, it just has a proton and an electron. Neutrons are funny little guys; sometimes they match the number of protons and sometimes they don't. I will tell you how many neutrons you need for each type of atom.**

neutron 2
proton 2
electron 2

2. **Helium is #2. (HEE-lee-em)**
   - **Put another blank piece of paper on the far right side of your work surface. Glue or tape the name "helium" onto the top of the paper.**
   - **You need two protons and two neutrons for helium. Did you put these in the center of the atom? They make up the nucleus.**
   - **Next, you need two electrons orbiting the nucleus.**

neutron 4
proton 3
electron 3

3. **Lithium is #3. (LITH-ee-em)**
   - **Start a second row of atoms by placing a blank piece of paper with the label for lithium below the hydrogen atom.**
   - **Lithium has three protons and four neutrons in its nucleus.**
   - **All the atoms you are making today have the same number of electrons as protons. How many electrons does lithium have? (3) Write the number of electrons on the label.**

Now I want to teach you something important. Answer these questions for me.
How many protons does hydrogen have? (1) How many electrons does hydrogen have? (1)
How many protons does helium have? (2) How many electrons does helium have? (2)
How many protons does lithium have? (3) How many electrons does lithium have? (3)
Do you see a pattern?
When going from one type of atom to the one that is next in line, you ALWAYS add 1 and ONLY 1 proton and 1 electron. Let's make more atoms and see how this works.

neutron 5
proton 4
electron 4

4. **Beryllium is #4. (beh-RIL-ee-em)**
   - **Place the paper for beryllium to the right of lithium.**
   - **If hydrogen has 1 proton, helium has 2 protons, and lithium has 3 protons, how many protons does beryllium have? (4)**
   - **Make the nucleus for beryllium. It has five neutrons.**
   - **Beryllium has four electrons.**
   - **Write the number of protons and electrons on the label.**

neutron 6
proton 5
electron 5

5. Boron is #5. (BO-ron)
- Put a sheet of paper and the label for boron next to beryllium.
    - Boron has six neutrons.
    - How many protons and electrons does it have? (5)
- Construct a marshmallow boron atom and write the number of protons and electrons on the label.

neutron 6
proton 6
electron 6

6. Carbon is #6. (KAR-ben)
- Put a sheet and the label for carbon next to boron.
    - A carbon atom has six neutrons.
    - How many protons and electrons does it have? Remember boron had 5. (5+1 = 6)
- Construct a marshmallow carbon atom and write the number of protons and electrons on the label.

Does it seem simple to make one type of atom and then the next? Well, it is. You are using marshmallows to make each type of atom. Every time you make an atom, you use the same kind of marshmallow for the protons and the same kind for the neutrons and the same kind for the electrons. If you could see something as small as a real atom, you would see that all electrons are the same as each other. All protons are the same as each other and all neutrons are the same as each other. What makes one type of atom different from another is the number of protons, neutrons, and electrons that it has. Amazing, isn't it?

neutron 7
proton 7
electron 7

7. Nitrogen is #7. (NYE-truh-gen)
- Put a sheet and the label for nitrogen next to carbon.
    - Nitrogen has seven neutrons.
    - How many protons and electrons does nitrogen have? Carbon had six of each. (7)
- Construct a marshmallow nitrogen atom and write the number of protons and electrons on the label.

neutron 8
proton 8
electron 8

8. Oxygen is #8. (OK-si-jen)
- Put a sheet and the label for oxygen next to nitrogen.
    - Oxygen has eight neutrons.
    - How many protons and electrons does oxygen have? Nitrogen had seven of each. (8)
- Construct a marshmallow oxygen atom and write the number of protons and electrons on the label.

neutron 10
proton 9
electron 9

9. Fluorine is #9. (FLOOR-een)
- Put a sheet and the label for fluorine next to oxygen.
    - Fluorine has ten neutrons.
    - How many protons and electrons does fluorine have? (9)
- Construct a marshmallow fluorine atom and write the number of protons and electrons on the label.

neutron 10
proton 10
electron 10

10. Neon is #10. (NEE-on)
- Put a sheet and the label for neon next to fluorine.
    - Neon has 10 neutrons.
    - How many protons and electrons does neon have? (10)
- Construct a marshmallow neon atom and write the number of protons and electrons on the label.

(continued on the back)

11. Move the helium atom in the first row over the top of the neon atom. (There should be a big space between hydrogen and helium.)

12. Weigh the marshmallow nucleus (protons and neutrons) of your neon atom on the scale. Now add the electrons for neon to the scale. Did the electrons make much of a difference to the overall weight? Where is almost all of the mass of the marshmallow neon atom? In real atoms (not just in marshmallow ones), almost all the mass is in the nucleus, too.

Now that you are an expert at making the different types of atoms, let's talk about energy levels. Leave the atoms you have made out; you will need them for this next section.

<u>Aloud Part 2 - Energy Levels:</u>
Can you find the energy level for helium on your "Parts of an Atom" poster? Remember, an atom's electrons go in its energy levels. How many electrons are in the energy level of helium on your poster? (2) Most atoms have more than one energy level. The energy levels have very strict rules about how many electrons can fit in each one. Think of it like musical chairs. There are only so many "seats" in each energy level, and each seat fits only one electron. Any more electrons in the atom have to go in the next energy level. The first energy level is the one closest to the nucleus. Only two electrons fit in this energy level. When the "music" stops at the first energy level, there can only be one or two electrons in it. The first energy level is always the first one to fill up.

Show the Atomic Energy Levels Diagram on the next page to your students, or recreate the diagram on a chalkboard.

Think of the nucleus of an atom as a planet and the electrons as moons. Hydrogen and helium have one energy level. Their electrons are like moons that have the same orbit. The atoms from lithium through neon have two energy levels.

13. Go back to your atoms. Put the electrons that go in the first energy level in their special "seats" in the energy level closest to the nucleus. Starting with hydrogen and ending with neon, put one to two electrons in the energy level closest to the nucleus.

That leaves most of the atoms with electrons that are not yet in an energy level. The second energy level can fit UP TO eight electrons in it. That is a lot of moons in one orbit!

14. Go to your marshmallow atoms and put all the remaining electrons in the second energy level around each nucleus. Do not bunch the electrons up; they should be spread evenly around the energy level.

Instructor's Note:

- Every type of atom has one and only one amount of protons. That is what defines the type of atom. The number of electrons can change as a function of bonding. In the neutral state, as represented on the periodic table and with your student's marshmallow atoms, the number of electrons equals the number of protons. The number of neutrons, however, can vary without changing the type of atom. In fact, every type of naturally occurring element has a variable amount of neutrons. These atoms are called isotopes. Isotopes are atoms with the same number of protons but different numbers of neutrons. Every type of atom has isotopes. Here and throughout this book, only the most commonly occurring number of neutrons is used.

Atomic Energy Levels Diagram

Maximum Number of Electrons

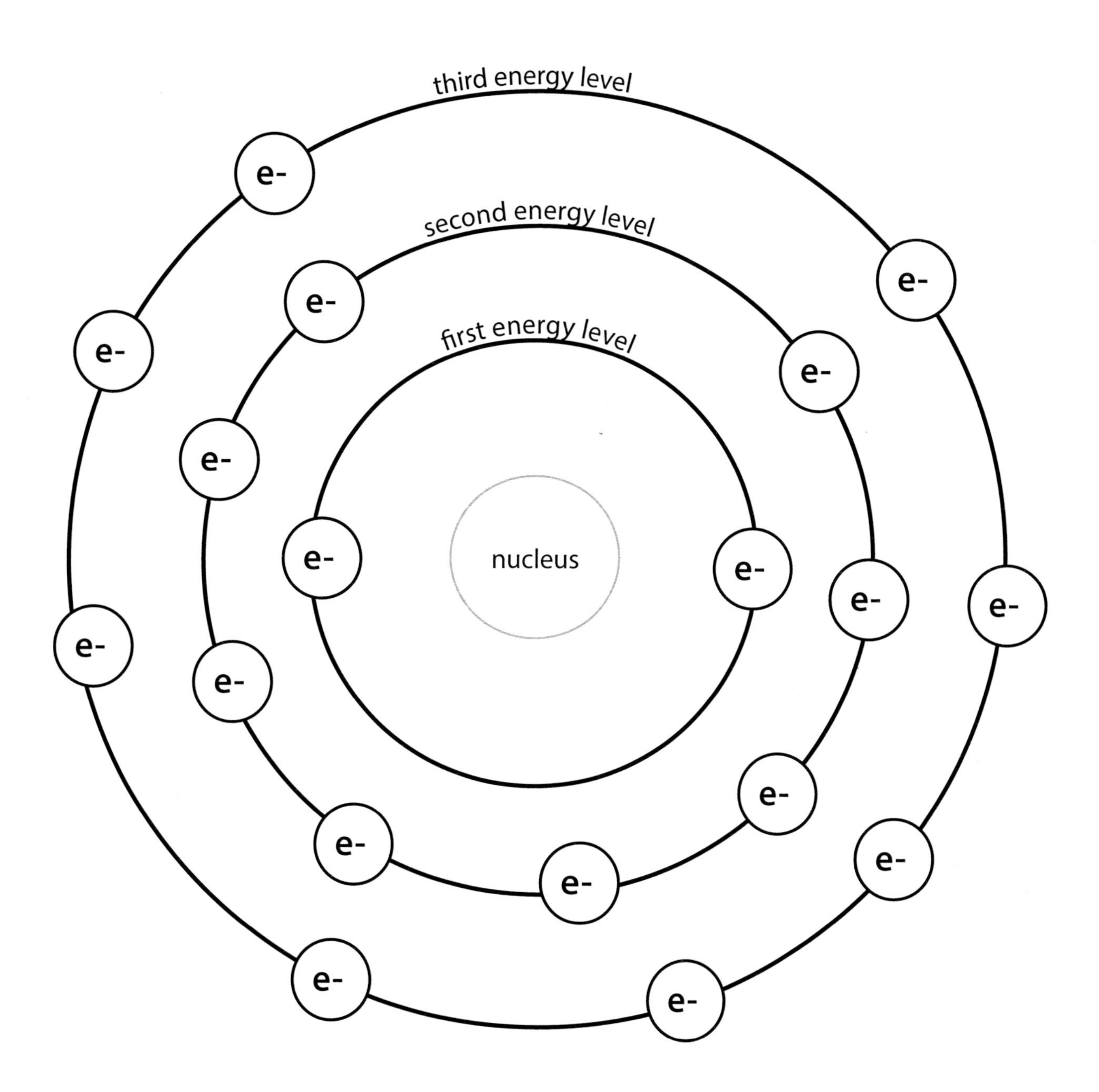

NAME ______________________________ DATE ________________

Types! Lab: The First Ten

Hydrogen 1 neutron 0 proton 1 electron 1	**Carbon 6** neutron 6 proton __ electron __
Helium 2 neutron 2 proton 2 electron 2	**Nitrogen 7** neutron 7 proton __ electron __
Lithium 3 neutron 4 proton 3 electron __	**Oxygen 8** neutron 8 proton __ electron __
Beryllium 4 neutron 5 proton __ electron __	**Fluorine 9** neutron 10 proton __ electron __
Boron 5 neutron 6 proton __ electron __	**Neon 10** neutron 10 proton __ electron __

Starting Small - Crossword Vocabulary Review

Across

4. A group of the same type of atoms.
7. Negatively charged particle that orbits the nucleus.
8. The center part of an atom, where the protons and neutrons are found.
9. When atoms join together, they make this.

Down

1. Where the electrons are found. (Two words)
2. The basic building block of all matter.
3. A positively charged particle found in the nucleus of an atom.
5. A neutral particle found in the nucleus.
6. The process where things mix without being stirred.

Unit 3

The Chemist's Alphabet Defined

The Periodic Table Rap

Each element in its own place.
Each one has its own special space.

Each type of atom has its own symbol.
Nicknames help them to be more nimble.

The periods are written in rows.
Left to right is how it goes.

Seven periods are we.
Just remember 4 + 3.

The groups are written up and down.
Chemical families like to hang around.

They are similar to each other,
These 18 groups are like brothers.

The number of protons equals the atomic number.
It is an important fact to "remumber."

Protons + neutrons is the atomic mass,
That's the fact I wrote down last

If you learn this you'll be great at chemistry.
Just you wait and see.

NAME ______________________ DATE ______________

For my notebook

The Alphabet

In 1869, a man named Dmitri Mendeleev (Men-de-LAY-ev) invented a chart to organize the different types of atoms. There are 118 different types, after all! Mendeleev's chart is called the Periodic Table of the Elements. The periodic table includes the name of every type of atom or element.

Atoms are the building blocks of the universe. Everything is made of them. The periodic table, with all the different types of atoms on it, is like the alphabet of chemistry. When reading and writing words, you use the alphabet from A to Z and make very simple words such as "I" or more complicated words such as "astonishment." Like letters in the alphabet, atoms can be found by themselves or they can combine to make something complicated, like a mountain. The elements are not on the periodic table in alphabetical order like the alphabet is, however. But you will soon learn that the order of the elements is very specific and very important.

All the names of the elements have abbreviations. These abbreviations are on the periodic table and they are called symbols. Look at the periodic table found in this book. Can you find a symbol for an element on the periodic table? Does the symbol have one, two, or three letters? There are a couple of rules for the symbols of elements: 1. Symbols have one, two, or three letters. 2. Like most abbreviations, the symbol is capitalized. 3. If there is more than one letter in the symbol, only the first letter is capitalized. Sometimes the symbols make sense. The symbol for hydrogen is **H**. Some symbols, though, do not seem to make sense. The symbol for potassium is **K**! Potassium is three squares below hydrogen on the periodic table. Can you find it? When a symbol does not make sense, it is because the symbol for that element comes from an earlier name we do not use anymore. The name for potassium used to be kalium.

So get ready to learn a whole new alphabet—the chemist's alphabet! But don't worry; this will probably be a lot easier than the first time you learned an alphabet.

The Alphabet Lab #1: My Periodic Table - instructions

Materials:
- "My Periodic Table" worksheets, 2 pages
- Pencil
- "Chemical Symbol Match" worksheet
- Periodic table found on the inside back cover of this book.

In this exercise, your student will be introduced to his periodic table. He will be adding to his periodic table in several future lessons. It will eventually become part of his "Element Book." He will only be making a couple of labels on it today. You might want to leave the periodic table worksheets in the book until the next unit. Or you can remove the table worksheets from this book, but do not cut, stable, assemble, or fold the table. The table will be cut, glued, and assembled in the next unit.

Aloud: Take out the "My Periodic Table" worksheets. First, your table needs a title. It is yours, so you get to name it. Your title should be written on the line directly above the large square at the top of the first page. After you have put a title on your table, find the big rectangle under your title. This rectangle will contain the key for your periodic table, just like a map key. You will start the key today and add to it later. You need to write in your neatest handwriting. Do not write too large, either. In the middle of the box, write the words "Element's Symbol." Keep your periodic table in a safe place; you will be working with it a lot more later.

Procedure:
When done titling the table, have your student complete the "Chemical Symbol Match" worksheet. Use the periodic table found on the inside back cover of this book for assistance. As students are matching the element names to the correct symbol, discuss the elements listed and their common uses.

Oxygen (O) - used in welding, water purification, and cement; required for supporting life and combustion
Gold (Au) - used to make, jewelry, art, coins; also used in dentistry
Nitrogen (N) - most of the air around us is nitrogen gas; used in fertilizers, liquid nitrogen for freezing
Potassium (K) - used to make glass, soap, lenses, and salt substitute. Also used to make purple fireworks!
Zinc (Zn) - used as a protective coating on steel; also used in paints, pennies, and rubber
Aluminum (Al) - foil, building products, most abundant metal found in Earth's crust
Chlorine (Cl) - used in water purification and bleaches
Iron (Fe) - most iron is used to make steel, which has many uses; powdered iron is used in magnets
Sulfur (S) - used in acids, fertilizers, and explosives
Silver (Ag) - used in jewelry, batteries, mirrors, electronics, and silverware
Carbon (C) - used in pencils, diamonds, steel, plastics, paint, and carbon dating
Lead (Pb) - used in gasoline tanks, lead batteries, and ceramics

My Periodic Table page 1

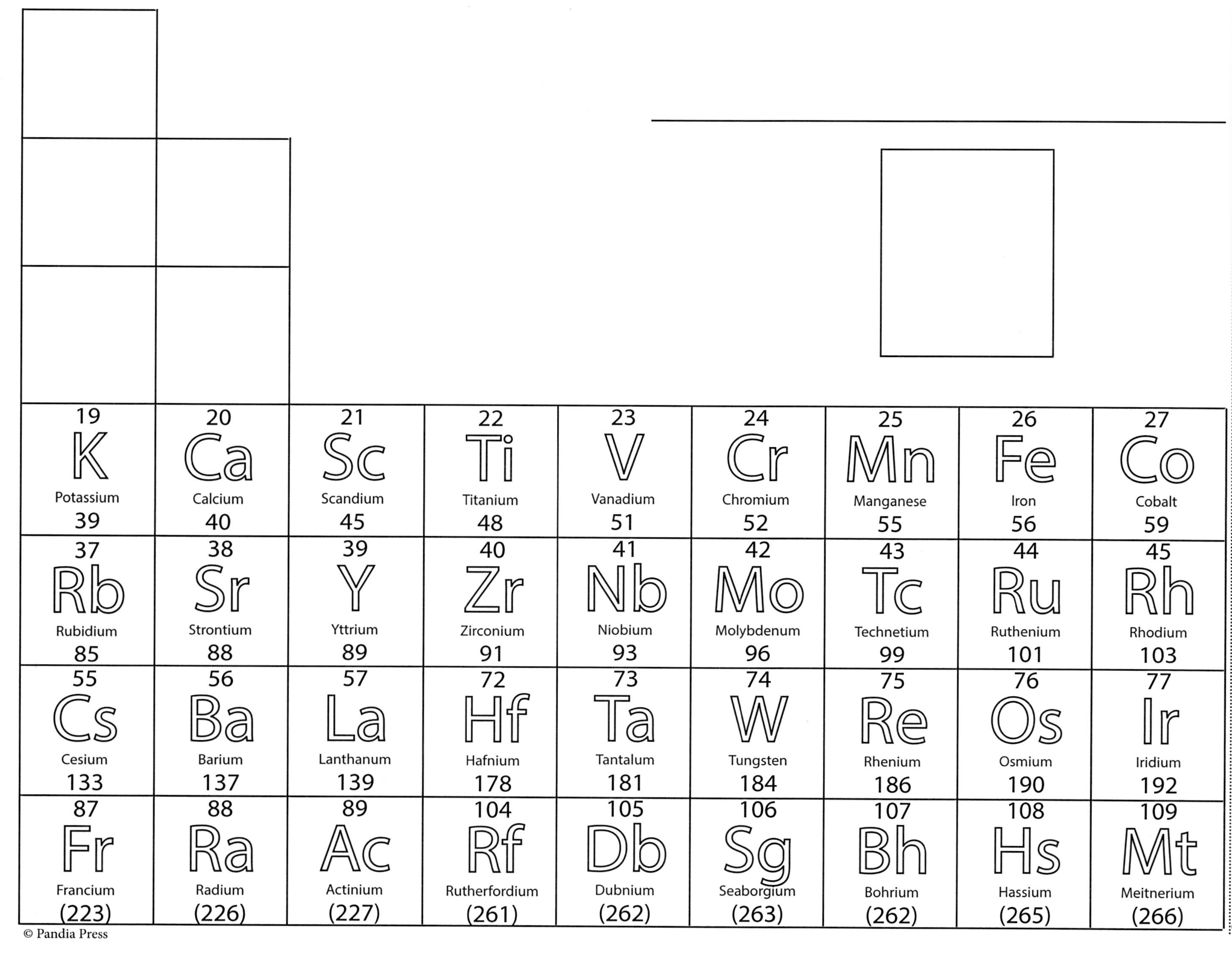

My Periodic Table page 2

28 Ni Nickel 59	29 Cu Copper 64	30 Zn Zinc 65	31 Ga Gallium 70	32 Ge Germanium 73	33 As Arsenic 75	34 Se Selenium 79	35 Br Bromine 80	36 Kr Krypton 84
46 Pd Palladium 106	47 Ag Silver 108	48 Cd Cadmium 112	49 In Indium 115	50 Sn Tin 119	51 Sb Antimony 122	52 Te Tellurium 128	53 I Iodine 127	54 Xe Xenon 131
78 Pt Platinum 195	79 Au Gold 197	80 Hg Mercury 201	81 Tl Thallium 204	82 Pb Lead 207	83 Bi Bismuth 209	84 Po Polonium (209)	85 At Astatine (210)	86 Rn Radon (222)
110 Ds Darmstadtium (281)	111 Rg Roentgenium (280)	112 Cn Copernicium (285)	113 Nh Nihonium (285)	114 Fl Flerovium (289)	115 Mc Moscovium (289)	116 Lv Livermorium (293)	117 Ts Tennessine (294)	118 Og Oganesson (294)

NAME ______________________________ DATE ____________________

Chemical Symbol Match

Match the name of each element with its symbol.

Oxygen	Pb
Gold	Fe
Nitrogen	Au
Potassium	Cl
Zinc	O
Aluminum	S
Chlorine	C
Iron	Ag
Sulfur	Zn
Silver	N
Carbon	K
Lead	Al

The Alphabet Lab #2: My Favorite Element – instructions

CAUTION: This lab might involve a heat source that can burn. Only the parent/instructor should use the heat source.

Materials:

- Lab sheet, pencil
- Lemon juice (fresh squeezed or from a bottle) in a dish
- Paintbrush or cotton swab
- Sunlight or a heat source, such as a light bulb or element of a stove
- Periodic table found on the inside cover of this book, to help choose a favorite element
- Pot holder glove if you use a heat source other than the sun

Aloud: Do you have a favorite element? If you do, keep what it is a secret for right now. If you do not have a favorite element yet, now is the time to choose one. Look at a periodic table and find the chemical symbol for your favorite element. Now write it down with invisible ink, which some people call "lemon juice," and see if someone else can figure out what your favorite element is.

Procedure:

1. Paint the symbol for your favorite element on the lab sheet.
2. Let the paper dry completely.
3. Hold the paper up to a heat source. If using a light bulb (this does work best), wear a glove pot holder and hold the paper near the bulb.
4. The writing will turn dark as it heats up. Don't let the paper touch the heat source, as it could catch on fire (which is another type of chemical reaction altogether).

Instructor's Note:

- A light bulb, the burner on your stove, or a hot wood-burning stove are the best heat sources to use for this experiment. These heat sources can burn, so be careful.

NAME ______________________________ DATE ____________________

The Alphabet Lab #2: My Favorite Element

NAME ______________________________ DATE ________________

For my notebook

Atomic Numbers

1	2	3	4	5
1 H 1 Hydrogen				
3 Li 7 Lithium	4 Be 9 Beryllium			
11 Na 23 Sodium	12 Mg 24 Magnesium			
19 K 39 Potassium	20 Ca 40 Calcium	21 Sc 45 Scandium	22 Ti 48 Titanium	23 V 51 Vanadium
37 Rb 85 Rubidium	38 Sr 88 Strontium	39 Y 89 Yttrium	40 Zr 91 Zirconium	41 Nb 93 Niobium

Look at the periodic table found inside the cover of this book. Do you see the number above each element's symbol? This number is called the atomic number. The atomic number of an element is a very helpful number because it ALWAYS equals the number of protons in the nucleus of the atom for that element. Therefore, an element is a group of atoms that all have the same number of protons in the nucleus. If hydrogen has an atomic number of 1, how many protons does it ALWAYS have in its nucleus? That's right, it has one proton. Do you remember from the marshmallow atoms that hydrogen has one proton in its nucleus?

Look at the periodic table again. Move two squares down below hydrogen to sodium. The elemental symbol for sodium is **Na**. What is the atomic number for sodium? If the atomic number of sodium is 11, how many protons does sodium ALWAYS have? That's right, it always has eleven protons. The atomic number also tells you how many electrons the atom has in its elemental form. Guess how many electrons sodium has? Maybe eleven is your lucky number. Without looking at the periodic table, I bet you can answer these questions:

How many protons does element #118 have?

How many electrons does it have in all its orbits?

Chlorine has 17 protons. What is the atomic number for chlorine?

How many electrons are orbiting around it?

You can tell a lot about an element from its atomic number. The atomic number is quite useful, don't you agree?

Now take out your periodic table (the one that has your name on it). Inside the key box write the words "Atomic Number" over the words "Element's Symbol." Put your periodic table back in a safe place; you're not done with it yet!

Atomic Number

Element's Symbol

Answers:

How many protons does element #118 have? (118)
How many electrons does it have in all its orbits? (118)
Chlorine has 17 protons. What is the atomic number for chlorine? (17)
How many electrons are orbiting around it? (17)

Atomic Numbers Lab #1: Is Milk an Element? - instructions

Materials:

- Lab sheet, pencil
- ½ cup Non-fat milk
- Glass
- 1 teaspoon Lemon juice
- Measuring cup

Aloud: Have you ever really thought about what is in milk? Is milk one thing or a mixture of things? Could milk be an element? It looks like one thing, doesn't it? How would you figure out the answer to a question like this one? If something is all one element, that means that all the atoms present are the same type. If you perform a chemical test on something that only has one type of atom, all parts of it will react the same way during the test. If you perform a test on milk and it separates into more than one part, then it cannot be an element.

Procedure:

1. Complete the hypothesis portion of the lab report.
2. Pour ½ cup of milk into a glass.
3. Stir 1 teaspoon of lemon juice into the milk.
4. Wait 20 minutes.
5. Dip your finger into the milk and see the coagulated bits of milk curd that have separated from the clear liquid whey.

Instructor's Notes:

- This very simple experiment might look familiar to you, if you like to bake. If you need buttermilk for a recipe and do not have any, this mixture works as a substitute. Baking powder in milk will work, too.
- When an acid, such as lemon juice, is added to a protein food, such as milk, the acid causes the protein to coagulate.
- Non-fat milk is not an element—it is a mixture of about 90% water, less than 0.5% milk fat, and 8.25% milk solids other than fat. Milk also has vitamin A and usually vitamin D added.

Possible Answers:

Hypotheses:

A – I think milk is an element because it looks like one thing.
B – I do not think milk is an element because it is not on the periodic table.
C – I do not know if milk is an element because I haven't done this experiment yet.

Results:

When lemon juice is added to the milk, the milk coagulates.

Conclusion/Discussion:

The correct answer is B. When lemon juice is added to the milk, causing it to coagulate, the milk has clearly been separated into parts.

NAME ______________________ DATE ______________

Atomic Number Lab #1: Is Milk an Element?

Hypothesis: Before you add lemon juice to the milk, answer either A, B, or C.

A. I think milk is an element because

____

____

____.

B. I don't think milk is an element because ____

____.

C. I don't know if milk is an element because ____

____.

Results: Draw a picture of what happened to the milk.

When I added lemon juice to milk, it looked like this:

Discussion/Conclusion: After you add the lemon juice, answer either A or B.

A. I think milk is an element because ____

____.

B. I don't think milk is an element because ____

____.

Atomic Number Lab #2: Flipbook - instructions

Materials:

- Flipbook activity sheets (three pages)
- Scissors
- Two different-colored pencils or pens
- Stapler
- "Atomic Energy Levels Diagram" found on page 59
- Periodic table found on the inside back cover of this book

Aloud: The atoms (elements) on the periodic table are arranged in order of their atomic numbers, the number of protons in the nucleus. Hydrogen, with an atomic number of 1, has one proton in its nucleus. It is found at the top left, the first spot, on the periodic table. Oganesson, with an atomic number of 118, has 118 protons in its nucleus. It is found at the bottom right, the last spot, on the periodic table. Today you will make a flipbook of elements number 1 to number 18. You will be drawing the protons in as you go. Remember, you add one and only one proton as you go from one element to the next. Also remember that the number of protons ALWAYS equals the atomic number.

You will also be drawing in the electrons. On the periodic table, the number of electrons equals the atomic number too. You have to pay attention to the energy levels as you draw the electrons in, though. Just remember, two electrons are all that will fit in the first (inner) energy level. Eight electrons are all that will fit into the second energy level, and eight electrons are the number of electrons in the third energy level found in nature for these elements.

Procedure:

Do all of the coloring and drawing before cutting the flipbook pages out. Refer to the "Atomic Energy Levels Diagram" on page 59 for help with the number of electrons in each energy level.

1. Draw the number of protons for each element.

   The names, atomic numbers, neutrons, and energy levels have already been drawn for each element. Put the protons in one at a time; they can be drawn on top of and next to the neutrons. The first element, hydrogen, with an atomic number of 1, will have one proton. The next element, helium, with an atomic number of 2, will have two protons, and so on. Use the same color for all the protons because every proton is the same, so they will all be the same color.

2. Using a different color, draw the electrons on the energy level of each element.

   Remember: The number of electrons EQUALS the number of protons. Put the electrons in one at a time. Electrons do not like to be right next to each other, so do not bunch them up around the energy level. The first element, hydrogen, will have one electron in it. The next element, helium, will have two electrons in it, and so on.

   Up to two electrons go in the first energy level, closest to the nucleus.

   ------Stop when you get to lithium------

3. From lithium to neon, there are two energy levels. The number of electrons is still the same as the number of protons. Draw two electrons in the first (inner) energy level. The rest of the electrons will go in the second energy level.

   Up to eight electrons go in the second energy level from the nucleus.

   -----Stop when you get to sodium------

4. From sodium to argon, there are three orbits. Draw two electrons in the first (inner) energy level. Next, draw eight electrons in the second orbit. The rest of the electrons, eight, will go in the third energy level.

(continued on the back)

5. Cut the flipbook along the lines on each page. Staple the pages together, starting with #1, hydrogen, and ending with #18, argon. You may need to cut the outside edge again. The pages need to be the same length on that end for the book to flip well from one page to the next.
6. Flip the book and watch the atoms change as the number of electrons, protons, and neutrons change.

Instructor's Note:

- This flipbook will be used as a reference later in the course.
- The third energy level will accommodate a maximum of 18 electrons. This is seen in the elements found lower on the periodic table.

Flipbook page 1

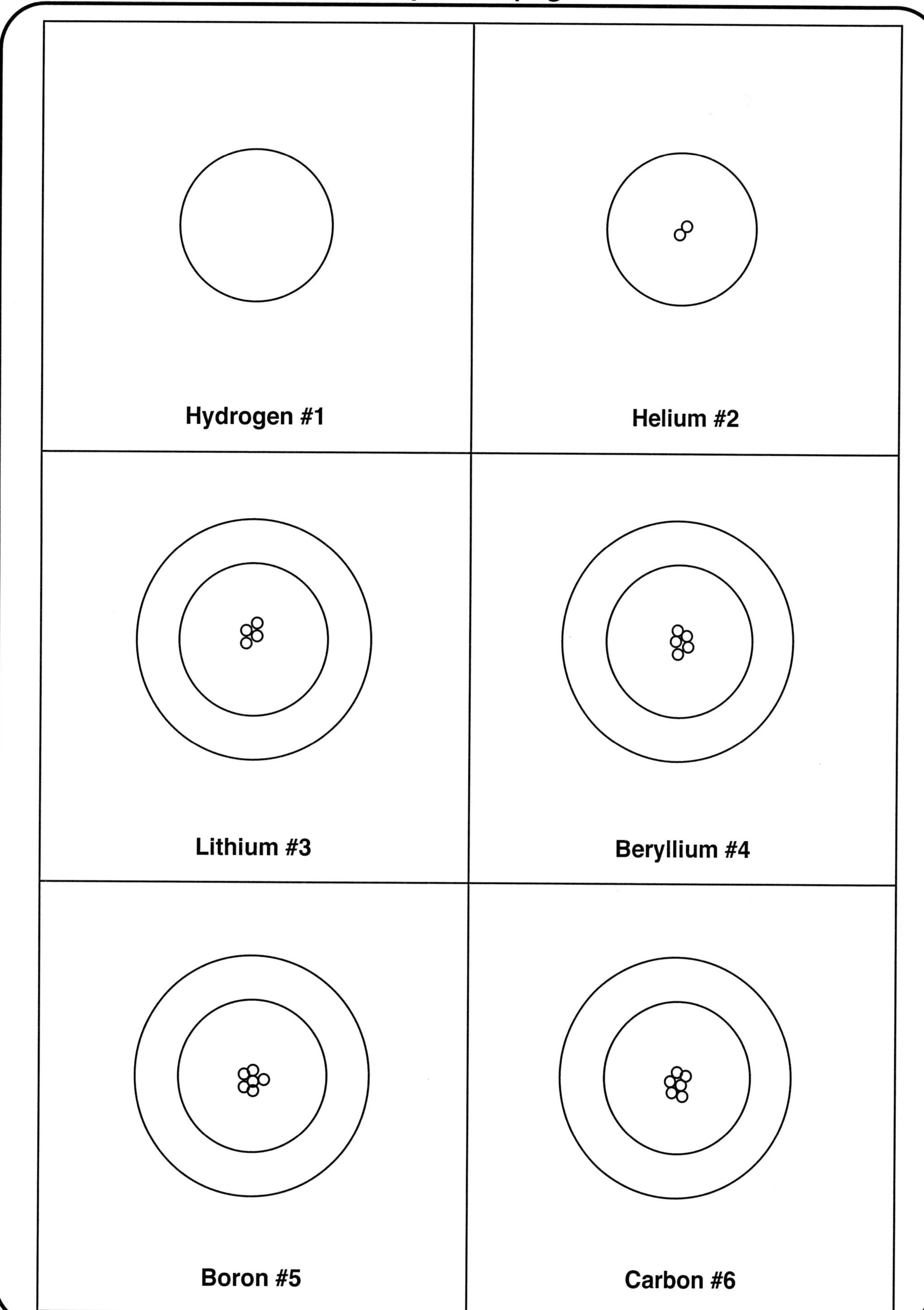

Flipbook page 2

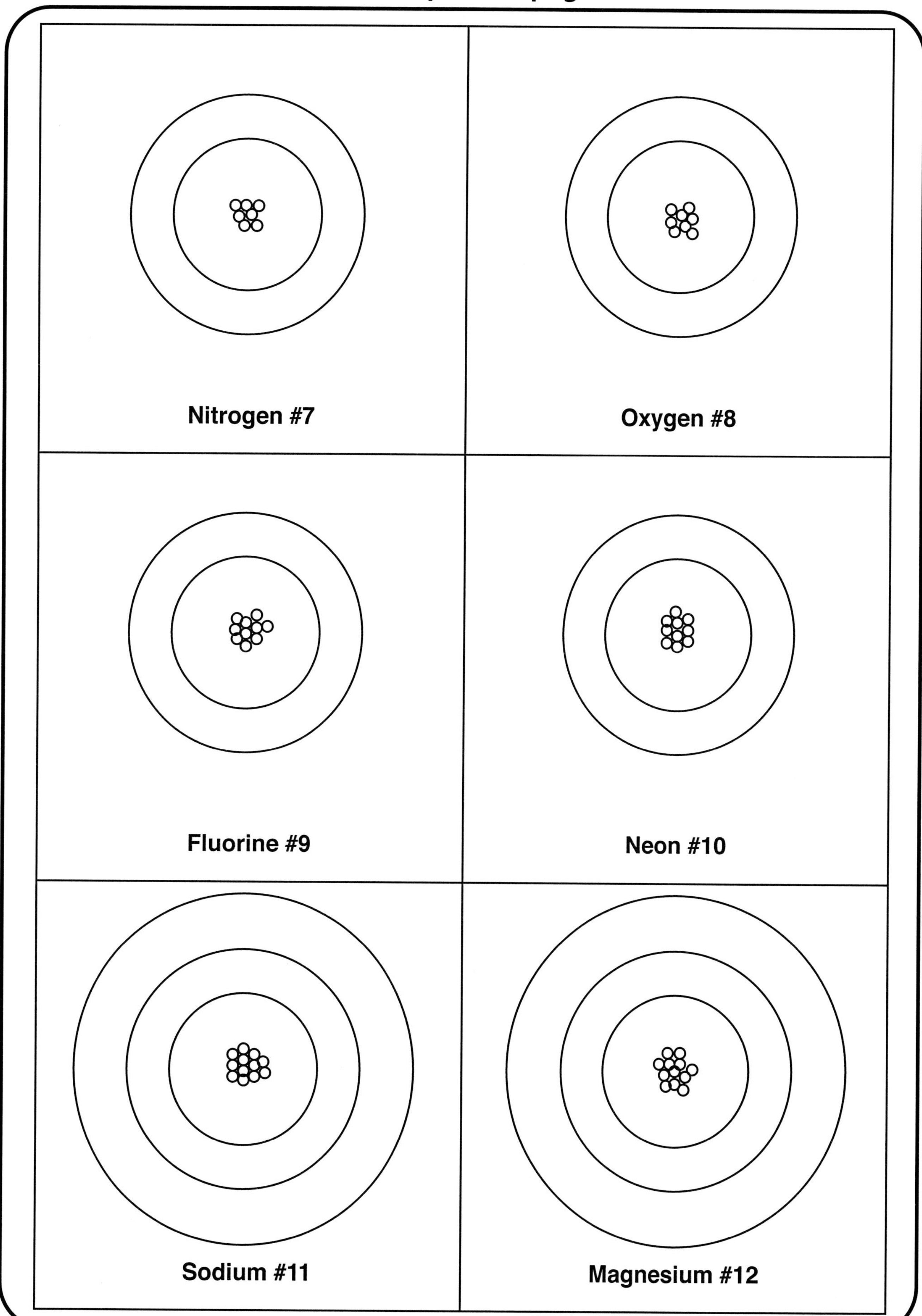

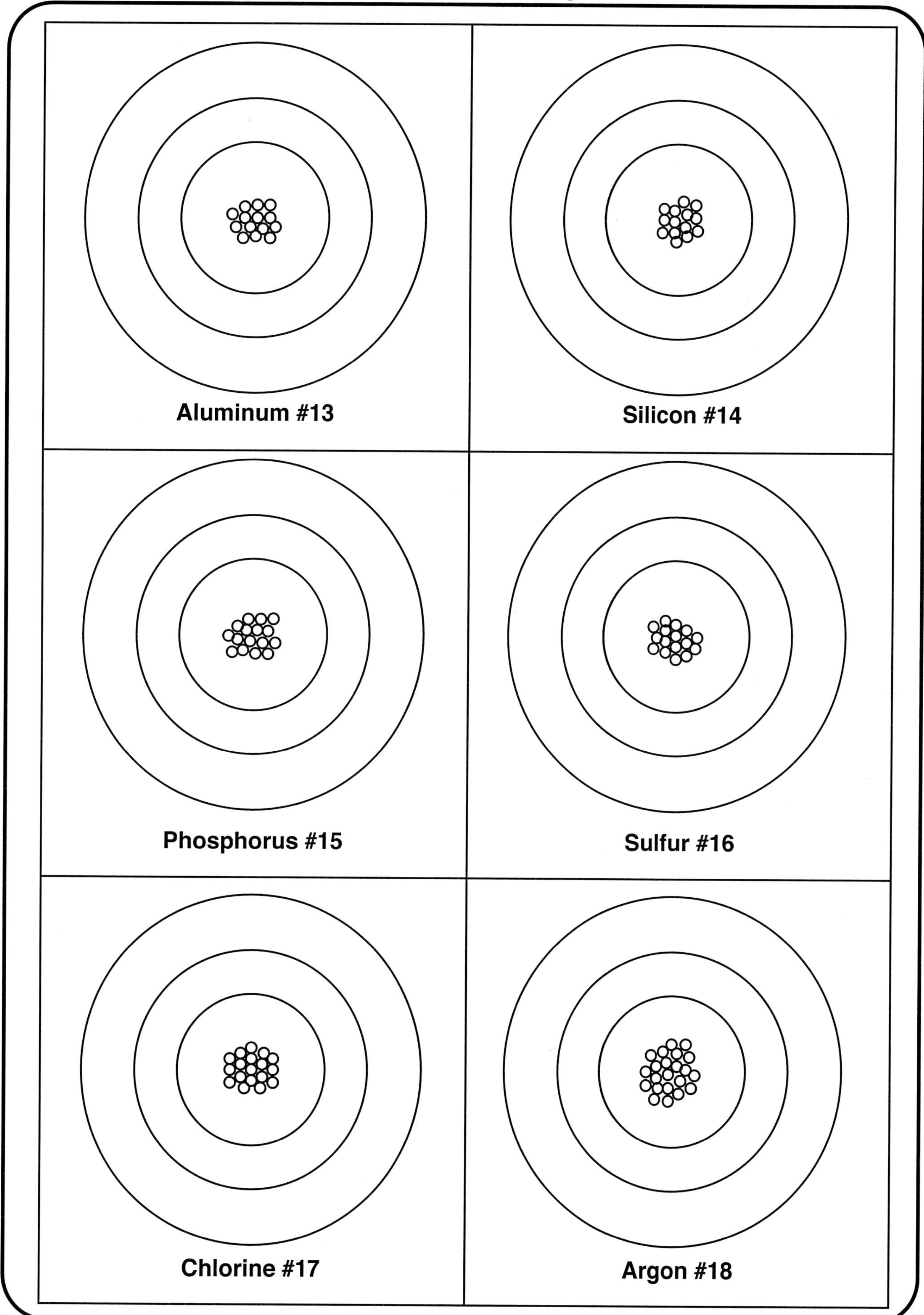
Aluminum #13
Silicon #14
Phosphorus #15
Sulfur #16
Chlorine #17
Argon #18

Pandia PRESS

NAME ______________________ DATE ______________

For my notebook

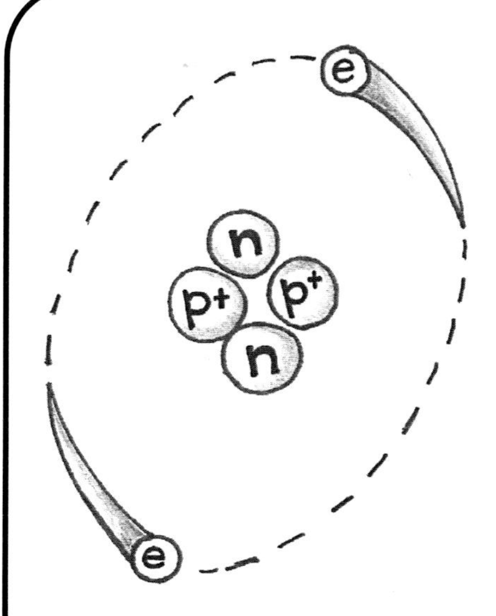

Massive Matters

Electrons are really small. Like almost all really small things, they weigh very little. Most of the mass of an atom is in the nucleus. Have you ever heard the term "mass" before? Mass is a lot like weight. The difference between mass and weight is that the mass of an object is the same anywhere in the universe. But weight is affected by gravity. So the weight of an object will be less on the moon than on Earth because the gravitational force of the moon is less than the gravitational force of Earth. But the mass will be the same. We talk about mass, not weight in chemistry, because that is what scientists do, but you won't notice a difference because on Earth, we can measure an object's mass by weighing it.

Most of the weight or mass of an atom is found inside the nucleus, so the mass of an atom is determined by adding the number of protons (the atomic number) to the number of neutrons. Scientists do not use grams or pounds for the mass of atoms; they use atomic mass units (a.m.u). The atomic mass units of atoms are on the periodic table.

Look at your "Parts of an Atom" poster. There are two protons and two neutrons in the nucleus of the helium atom, so that means that helium would have an atomic mass of 2 protons + 2 neutrons, or 4 a.m.u. Now find the symbol for helium, He, on the top right corner of the periodic table. Do you see the number below the symbol? That is the atomic mass of helium and it is 4.

When you are calculating the atomic mass, the number of protons is easy to figure out, because it is the atomic number. The number of neutrons is not as easy to figure out. It would be nice if the number of neutrons matched the number of protons, but a lot of the time it does not. The number of neutrons can vary, so in this book I will always tell you the number of neutrons, and you can figure out the rest.

Take out your periodic table. In the box on top, below the words "Element's Symbol," write the words "Atomic Mass." That is where you will write the atomic mass of the elements as you fill in your periodic table.

Atomic Number
Element's Symbol
Atomic Mass

Massive Matters Lab #1: My Favorite Element Explored - instructions

Materials:

- Copy of lab sheet, pencil
- Internet access and/or chemistry books/encyclopedias
- Colored pencils or crayons
- Periodic table found in the inside cover of this book

Aloud: Choose a favorite element from the periodic table. You can choose the same element you wrote in secret ink or maybe you have a new favorite element. Look up your favorite element on the Internet or in an encyclopedia. Go to google.com, type in the name of the element, and see what comes up. You can find out how your element is used, when it was discovered, and for what or for whom it was named.

NAME ______________________________ DATE ________________

Massive Matters Lab #1: My Favorite Element Explored

What is the name of your favorite element?

____

What are the atomic number, atomic mass, and symbol of your favorite element?

__________ __________ __________

Why is it your favorite element?

Draw a picture of one atom of your favorite element:

Write down four facts you learned about your favorite element:

1. ____

____.

2. ____

____.

3. ____

____.

4. ____

____.

Massive Matters Lab #2: Which Weighs More? - instructions

A sealable baggie with 1 cup of water needs to be frozen before this experiment.

Materials:

- Lab sheet, pencil
- 1 cup Water, frozen in a baggie
- 1 cup Liquid water
- 1 cup Powdered sugar
- 1 cup Brown sugar
- 1 cup of Grapes
- 1 cup Grape juice
- Kitchen scale that measures grams
- One-cup measuring cup
- Six sealable baggies

Aloud: Have you ever picked something up and found it was a lot heavier than you thought it was going to be? Which would have a larger mass, a big box of cereal or a gallon of milk? They take up about the same amount of space, though, don't they? The relationship between mass and the space something takes up is a property chemists can use to tell different things apart.

For this experiment, you are going to compare three different sets of things. First, you will predict which has a larger mass. Then you will measure the masses by weighing the items on a scale to find out if you were correct.

Procedure:

1. Freeze 1 cup of water in a baggie overnight.
2. The next day, pour 1 cup of liquid water into a baggie.
3. Measure 1 cup of powdered sugar and pour it into a baggie.
4. Measure 1 cup of packed brown sugar and pour it into a baggie.
5. Measure 1 cup of grapes and pour them into a baggie.
6. Measure 1 cup of grape juice and pour it into a baggie.
7. Complete the hypothesis portion of the lab sheet
8. Weigh each baggie on the scale. Record the actual masses in grams.

Aloud: Why is it that things that take up the same amount of space do not always have the same mass? Sometimes it is because of how the things fill the space. For instance, grapes do not fill all the space in the measuring cup; there is air between the grapes. The grape juice does fill the entire space of the cup. The mass of a cup of air is less than a cup of grapes or grape juice. Sometimes the difference is because of the amounts and types of atoms present in what you are weighing. As you can see from looking at a periodic table, there is a big difference between the atomic mass of hydrogen (H), which is 1, and that of gold (Au), which is 197. A cup of hydrogen would weigh a lot less than a cup of gold because its atomic mass is a lot less.

Did you guess the water would have the same mass (weigh the same) as the ice? It does because there is the same amount of the same kind of molecules in both baggies. Freezing the water molecules does not change the mass. It does affect which one you would rather have drop on your foot, though!

(continued on the back)

Instructor's Note:

- The distinction between mass and weight is one that stumps many science students (young and old). Mass is usually measured in grams or kilograms and is most accurately measured on a balance scale, like a triple beam scale. For this age group, however, we will measure mass by measuring weight on a common scale. When you measure weight, the scale shows only an approximate value of mass for a given weight.

NAME ______________________________ DATE ________________

Massive Matters Lab #2: Which Weighs More?

In the table below, make predictions by writing the names of the items you think have the larger mass. If you think they weigh the same, write an S for "same."
Then weigh the items and see if your prediction was accurate.

	Ice/Water	P. Sugar/B. Sugar	Grapes/Grape Juice
Prediction			
Actual Weight/ Mass	Ice =	P. Sugar =	Grapes =
	Water =	B. Sugar =	Grape Juice =

P. = powdered B. = brown

My Predictions and the Actual Facts

❒ I am so good at predicting masses that I must have a scale in my head.

❒ I am so bad at predicting masses I could not tell a watermelon from a cherry.

I was surprised that ______________ weighed (more) (the same) (less) than/as

______________________.

1 cup of ________________ had the smallest mass of the six things I weighed.

1 cup of ________________ had the largest mass of the six things I weighed.

NAME ______________________ DATE ______________

For my notebook

Why Do They Call It the PERIODic Table Anyway?

Look at the periodic table found inside the cover of this book. The rows of the periodic table have a special name. They are called periods. The periods on the periodic table have the same name as periods that go at the end of sentences, but it doesn't mean the same thing. Starting with hydrogen, H, count down three elements; your finger should be on potassium, K. Now take your finger and trace it across the fourth row. The fourth row is called the 4th period. This row starts with the element potassium, K, and ends with the element krypton, Kr. How many periods are on your periodic table? Let's count the rows and find out. There are seven rows on the periodic table. That means there are seven periods on the periodic table. Do all the periods have the same number of elements in them? (No) What is the shortest period? (The 1st period, with H and He in it.)

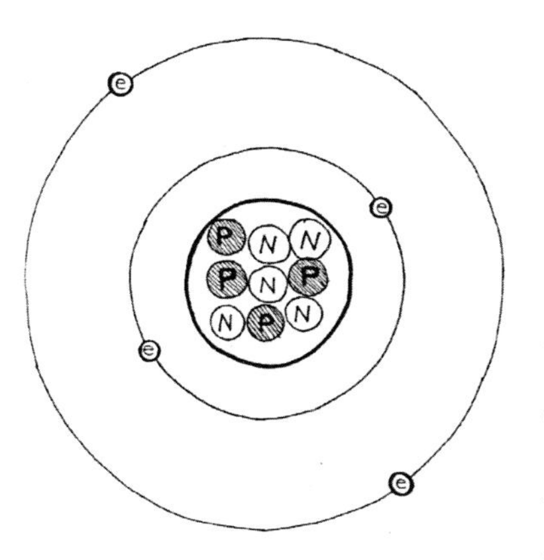

The period an element is in tells you the number of energy levels that element has. Electrons move around all the time. All that movement creates energy. Elements in the same period have the same number of energy levels. For example, all elements in the 4th period have four energy levels.

Now locate your periodic table. On the title of your periodic table, underline the word "period" found in the word Periodic. In small-sized numbers, write the ordinal number for each period from 1st to 7th on the left side of your table. The first row is on the top and the seventh row is on the bottom. Turn your periodic table a ¼ turn clockwise. Write the word PERIODS beside the numbers.

PERIODS	
1st	H
2nd	Li
3rd	Na
4th	K
5th	Rb
6th	Cs
7th	Fr

Why Do They Call It the PERIODic Table Anyway?
Worksheet- instructions

Materials:

- Worksheet, pencil
- Periodic table found on the inside back cover of this book

Procedure:

Use the periodic table to answer the questions on the lab worksheet.

Answers:

Energy Levels Quiz

How many energy levels are in the 1st period? 1

How many energy levels are in the 2nd period? 2

How many energy levels are in the 3rd period? 3

How many energy levels are in the 4th period? 4

How many energy levels are in the 5th period? 5

How many energy levels are in the 6th period? 6

How many energy levels are in the 7th period? 7

The Metals Quiz

What period is tin, Sn (50)*, in? 5th

What period is gold, Au (79), in? 6th

What period is silver, Ag (47), in? 5th

What period is aluminum, Al (13), in? 3rd

What period is copper, Cu (29), in? 4th

What period is mercury, Hg (80), in? 6th

Instructor's Note:

*The atomic number is given in parenthesis after the elemental symbol for each metal element to help you locate it on the periodic table.

NAME ______________________________ DATE ________________

Why Do They Call It the PERIODic Table Anyway? Worksheet

ENERGY LEVELS QUIZ

How many energy levels are in the 1st period? ____________________

How many energy levels are in the 2nd period? ____________________

How many energy levels are in the 3rd period? ____________________

How many energy levels are in the 4th period? ____________________

How many energy levels are in the 5th period? ____________________

How many energy levels are in the 6th period? ____________________

How many energy levels are in the 7th period? ____________________

THE METALS QUIZ

What period is tin, Sn (50), in? ____________________

What period is gold, Au (79), in? ____________________

What period is silver, Ag (47), in? ____________________

What period is aluminum, Al (13), in? ____________________

What period is copper, Cu (29), in? ____________________

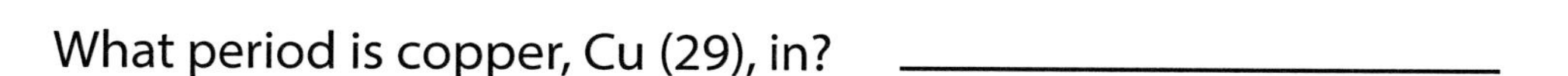

What period is mercury, Hg (80), in? ____________________

Why Do They Call It the PERIODic Table Anyway?
Lab: Periodic Play Dough - instructions

Materials:

- 1 cup Peanut butter*
- ¼ cup Honey
- 1 cup Dry powdered milk
- One-gallon-size sealable baggie
- Mixing bowl
- Measuring cup
- Wooden spoon
- Flipbook for reference
- Periodic table for reference

There is not a lab sheet for this lab.

Aloud: Today you are going to make periodic play dough. You can use it to sculpt things, and it tastes yummy! When you make sculptures with it, you will feel little particles in it. Pretend the particles are atoms of an element. Which period is the element in? Form the play dough into the shape of the symbol for the element, the period number it is found in, and the shape of one atom in the element with protons, neutrons, electrons, and energy levels.

Procedure:

1. Mix the first three ingredients in the mixing bowl with the wooden spoon until well mixed.
2. The powdered milk particles or "atoms" can be felt in the play dough. Students can use their flipbooks and periodic tables for reference to mold the play dough into shapes of atoms, period numbers, and/or elemental symbols. The play dough is edible and delicious!
3. Leftovers are perishable, so store in a sealed container and refrigerate.

Instructor's Notes

- If the dough is sticky, try adding a little more powdered milk, or add a little flour to the mixture.
- This recipe makes about 2 cups of play dough. It can be doubled or tripled if making play dough for a crowd.
- * ½ cup of cream cheese can be substituted for the cup of peanut butter in this recipe.

NAME ______________________________ DATE ________________

For my notebook

We Are Family

Your family group is important to you. The elements on the periodic table are in family groups, too. Put your finger on hydrogen, H. Now trace your finger all the way down this column. Did you end on francium, Fr? You have just traced your finger down Group 1 of the periodic table. Each column on the periodic table is a chemical family or group. There are 18 chemical families.

The elements in a chemical family have similar traits or properties. They also have things that are unique just to them. It is like your own family. Each member of your family has things that make them special. You might also share traits, such as the color of your hair or eyes, with the other members of your family.

Look at hydrogen's chemical family, Group 1. When you look at that collection of names and numbers and symbols, do you wonder what they have in common? It isn't their names or symbols, is it? Do you see any pattern for their atomic numbers: 1, 3, 11, 19, 37, 55, and 87? (No) Do you see any pattern for their atomic masses: 1, 7, 23, 39, 85, 133, and 223? (No)

What is it that makes this group of elements a chemical family? Take your flipbook out and let's look for a pattern. Hydrogen, lithium, and sodium are all Group 1 elements and they are all in your flipbook. Look at hydrogen. How many electrons does hydrogen have in its outer energy level? (1) Now look at lithium. How many electrons does lithium have in its outer energy level? (1) Finally, look at sodium. How many electrons does it have in its outer energy level? (1) The elements in Group 1 ALL have one electron in the outer energy level.

This means that all the elements of a group have similar traits to all the other elements in their group, because they have the same number of electrons in their outer energy level. But they are also unique, just like the members of your family.

We Are Family:
The Friendship of Beryllium and Boron - Worksheet instructions

Materials:
- "My Periodic Table"
- Worksheet, pencil
- Periodic table found in the inside cover of this book.

Procedure:
1. In the middle of your periodic table, above the squares, in that low spot, draw an arrow going across. Write the word "Groups" above this arrow. Starting with the number 1 and ending with the number 18, write the number of each group above the columns on your periodic table.

Groups →

1	2	3	4	5	6	7	8	9	10	11	12	13	14	15	16	17	18

2. Complete the fill-in-the-blank worksheet. Use the periodic table as a reference.

Answer Key:

Beryllium and Boron are best friends. They are even next-door neighbors. They live on a street called the 2nd row. In a town called Periodic Table. They have nicknames that are almost the same, too. Their nicknames are Be and B.

The town they live in, Periodic Table, is small. There are only 18 family groups in the entire town. Beryllium sometimes tells people he is Boron's twin brother, but they don't believe him because he looks and acts so much like the rest of his family, called Group 2. Besides, Beryllium's family is pretty well known in town; they like to get together and help make fireworks. Boron's family name is Group 13. Beryllium has 6 members in his family, and Boron also has 6 members in his.

They were so excited when they were picked to be on the same baseball team. The day the coach gave out numbers, Beryllium got 4 and Boron got 5. Beryllium wanted the larger number, but Boron got it by one number. Then Beryllium remembered his nickname was one letter longer, so maybe it was okay.

Beryllium and Boron both have brothers who live two streets down from them on a street called the 4th row. Their brothers also have nicknames that are nearly the same. Their brothers' full names are Calcium and Gallium, but everyone calls them by their nicknames, Ca and Ga. Their brothers are famous baseball players. Their numbers are 20 and 31. When Beryllium thought about it, his brother's number was 11 smaller than Gallium's. Maybe one number smaller is not so bad after all.

NAME ______________________________ DATE ________________

We Are Family: The Friendship of Beryllium and Boron

Fill in the blanks. Everything you need for the answers can be found on a periodic table.

Beryllium and Boron are best friends. They are even next-door neighbors. They live on a street called the ________ row. In a town called ______________ Table. They have nicknames that are almost the same, too. Their nicknames are _______ and ________.

The town they live in, Periodic ______________, is small. There are only _______ family groups in the entire town. Beryllium sometimes tells people he is Boron's twin brother, but nobody believes him because he looks and acts so much like the rest of his family, called Group _______. Besides, Beryllium's family is pretty well known in town; they like to get together and help make fireworks. Boron's family name is ___________13. Beryllium has ________ members in his family, and Boron also has ________ members in his.

They were so excited when they were picked to be on the same baseball team. The day the coach gave out numbers, Beryllium got ________ and Boron got _________. Beryllium wanted the larger number, but Boron got it by one number. Then Beryllium remembered his nickname was one letter longer, so maybe it was okay.

Beryllium and Boron both have brothers who live two streets down from them on a street called the 4th __________. Their brothers also have nicknames that are nearly the same. Their brothers' full names are ______________________ and ______________________________, but everyone calls them by their nicknames, _________ and _________. Their brothers are famous baseball players. Their numbers are _________ and _________. When Beryllium thought about it, his brother's number was 11 smaller than Gallium's. Maybe one number smaller is not so bad after all.

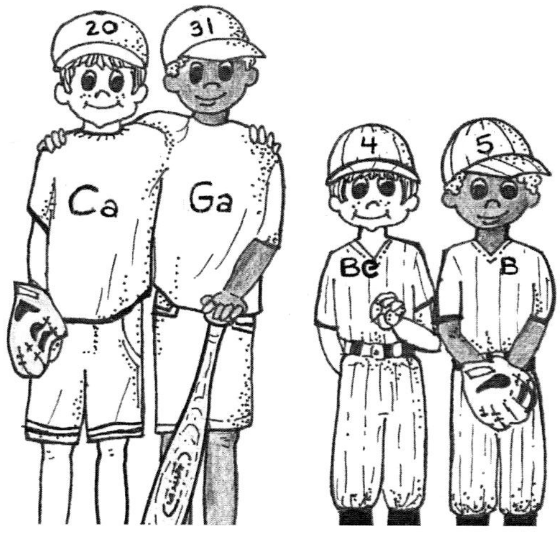

We Are Family Lab: Prove It! – instructions

Materials:

- Lab sheets (two pages), pencil
- ¼ cup Table salt (NaCl)
- ¼ cup Salt substitute (KCl - potassium chloride)
- 2 tablespoons Oil
- 2 tablespoons Vinegar
- One cooked potato cut in 2 halves (optional for tasting salt)
- 2 cups Water + 2 tablespoons
- 2 cups Crushed ice
- Two glasses
- Two small cups to hold both salts
- Tablespoon
- ⅛ teaspoon Measuring spoon
- Towel to dry glasses
- Thermometer (science-type or kitchen thermometer that will go down without being shaken)
- Internet access

Aloud: Do you ever wonder if something is a fact or an opinion? Could you ever be called a doubter, wondering if someone really knows what he or she is talking about? If you answered yes to either of these questions, then you have the makings of a good scientist! Scientists are constantly asking questions such as: What if? How can they be sure of that? Have they ever seen that happen or find any evidence, or did they just read about it in a book?

I told you that elements in a group have similar traits. They belong in the same family. How can you be sure that is true? Let's experiment and prove it, that's how. Look at a periodic table. In the first group, there is sodium and potassium. Both are used to make salt. Sodium is used to make table salt, and potassium is used as a salt substitute. They are in the same chemical family, so if we experiment with them, the results should be about the same. You will perform physical and chemical tests on both salts.

Procedure:

1. Complete the hypothesis portion of the lab sheet.
2. Measure each salt into a dish. Smell, look at, and touch the salts. They can be tasted as is or sprinkled on a potato half. Record observations.
3. Begin chemical tests by measuring ⅛ teaspoon of each salt into separate glasses. Add 1 tablespoon of water to each glass. Wait 5 minutes. Record observations. Clean the glasses out and dry them. Repeat this for vinegar and then oil.
4. Fill each empty, clean glass with crushed ice. Pour water into glasses, just to the top of the ice level. Put the thermometer in each glass for 3 minutes. They should both be the same temperature. Record the temperatures.
5. Add 2 tablespoons of table salt to one of the glasses. Put the thermometer in and stir the mix gently with the thermometer. Do this for 3 minutes. Record the temperature. Repeat this in the other glass with salt substitute.
6. Preview Web videos (for appropriateness) to observe what happens when pure sodium and potassium metal are dropped into water. Enter "pure sodium metal in water video" and "potassium metal in water video" into a search engine. Once you find appropriate videos, share them with your student (who will draw a picture of the reactions on the lab sheet). (You might want to revisit these video demonstrations after students learn the terms "indicator," "basic," and "exothermic" while studying chemical reactions in unit 7.)

(continued on the back)

Instructor's Notes:

- Sodium and potassium are both very reactive metals. They are so reactive that it is rare to find them in their elemental form. They are much more commonly found as salts. The Web videos show what happens when they are in their elemental form and come in contact with water.
- I found the taste of the potassium chloride to be salty but different from what I was used to. I would suggest comparing how the two salts taste on a cooked potato. It is more palatable that way.
- Most vinegar is 95% water, so expect the salts to dissolve the same way they do in water. The reason for doing the vinegar part of the experiment is to see how each salt reacts in an acid; vinegar is an acid. Remember how different baking powder behaved in water and vinegar.
- Hold the thermometer at a middle depth in the glass when you are reading the exact measurement. I found the bottom was slightly warmer than the middle.

Possible Answers:

Results/Observation:

The two salts behave the same in every physical test except taste. They do taste different. Not a lot different, but not the same, either.

Most table salt has additives. These may make the table salt solution a little cloudy when it dissolved in water and vinegar. The potassium salt will not become cloudy.

The temperature will fall when either salt is mixed into the ice water.

The drawings should show an explosion for both sodium and potassium.

Discussion/Conclusion:

The results from the physical tests and the chemical tests show that sodium and potassium behave about the same. They do have many traits in common. Therefore, they must be family.

NAME ______________________________ DATE ________________

We Are Family Lab: Prove It! - page 1

Group 1

H
Li
Na
K
Rb
Cs
Fr

Hypothesis: (circle your answer)

Sodium, Na, and potassium, K, are both in Group 1, which is one chemical family. Do you think sodium and potassium are going to behave about the same when you perform physical and chemical tests on them?

Yes No I don't know

Results/Observations:

Physical Tests: How well do these two compare?
Write S if they are about the same.
Write D if they are very different.
Write S/D if they are sort of different and sort of the same.

	Feel	Smell	Appearance	Taste
Sodium				
Potassium				

Chemical Tests: How well do these two compare?
Write S if they are about the same.
Write D if they are very different.
Write S/D if they are sort of different and sort of the same.

	Water	Vinegar	Oil
Sodium			
Potassium			

We Are Family Lab: Prove It! - page 2

Temperature:

	No salt	With sodium	With potassium
Ice water 1			
Ice water 2			

Draw pictures of what you saw happen on the Internet when elemental sodium and potassium were dropped into water.

Sodium	Potassium

Discussion and Conclusion:
How did sodium and potassium compare to each other? Did your results show that they are family or not?

NAME ______________________________ DATE ________________

For my notebook

It's ELEMENTary

The number of elements found on the periodic table changes from time to time because there are manufactured elements. At this time there are 118 different elements. The elements from number 93 to number 118 are manufactured. In a laboratory, scientists can sometimes change one of the naturally occurring elements to make a new one, a manufactured one. When scientists make a new element, the number of elements increases on the periodic table. Maybe you will become a chemist and create a new element one day. What will you name it if you do?

	1	2	3	4	5	6	7	8	9	10	11	12	13	14	15	16	17	18
1st	H																	He
2nd	Li	Be											B	C	N	O	F	Ne
3rd	Na	Mg											Al	Si	P	S	Cl	Ar
4th																		

Table of 18 commonly found elements

Most things, though, are made from the 18 naturally occurring elements on the table above. Some of these elements you have probably never heard of, like magnesium (Mg). One of the naturally occurring elements shown above is also sometimes manufactured. Can you guess what it is? Hint: It can be expensive to buy. Did you know that one of the elemental forms of carbon (C) is diamond?

Did you notice the numbers above each column or group? Remember, these are the group or chemical family numbers. You are going to be learning about the 18 commonly found elements, and these elements are in Groups 1 and 2 and Groups 13 through 18.

The table above might look like a random collection of letters to you. That is because you do not know much about these elements yet. But soon you might become somewhat of an "element expert."

It's ELEMENTary: Twenty Questions - instructions

Materials:

- Worksheet, pencil
- Periodic table (found on the inside back cover of this book)

Procedure:

Use a periodic table to answer to the questions.

Answers:

What is the first element in Group 14? Carbon

What is the one-letter abbreviation for element number 8? O

What is the full name of element 8? Oxygen

What is the atomic number of the last element in the 5th period? 54

Which comes first on the periodic table: nitrogen, N, or chlorine, Cl? Nitrogen

Group these elements by family. F Ti Br Zr Hf I
Family Group 4 - Ti, Zr, Hf
Family Group 17 - F, Br, I

How many periods are on the periodic table? 7

What is the name of the bottom element in Group 1? Francium

Li is the abbreviation for what element? Lithium

What element comes after neon, Ne? Sodium, Na, 11

Some of the elements have abbreviations that spell short one- or two-letter words. What are they?
He - #2, Be - #4, As - #33, In - #49, I - #53, and At - #85. Also from the "f-block" of inner transition metals (not pictured on your periodic table): Am - #95 and No - #102

Nickels are made from the element nickel. Give its atomic number and abbreviation. Ni, 28

Calcium is found in milk and is important for bones and teeth. Give its atomic number and abbreviation.
Ca, 20

Which period has the fewest elements in it? The 1st period, with only 2 - helium and hydrogen

If carbon has six neutrons and six protons, what is its atomic weight? $6 + 6 = 12$

How many electrons does carbon have? 6

What element has the abbreviation Co? Cobalt, 27

What element is the same number as your age? Answers will vary.

What element is the same number as your mom or teacher's age? True answers will vary, or just say copper (29).

Find Sulfur, S, on the chart. What element is next to it on the left? Phosphorus, P, 15

Scoring:

20 - 18 correct - You are a master of the periodic table.
17 - 15 correct - You have the makings of a good chemist.
14 - 12 correct - Better going next time.
11 or fewer correct - Next time use the periodic table right side up.

NAME ______________________________ DATE ____________________

It's ELEMENTary: Twenty Questions

Use a Periodic Table to answer these 20 questions.

1. What is the first element in Group 14? ____________________
2. What is the one-letter abbreviation for element number 8? ____________
3. What is the full name of element 8?____________________
4. What is the atomic number of the last element in the 5th period?__________
5. Which comes first on the periodic table: nitrogen, N or chlorine, Cl? ________
6. Group these elements by family. F Ti Br Zr Hf I

   Family Group 4 ____________________

   Family Group 17 ____________________
7. How many periods are on the periodic table? _______
8. What is the name of the bottom element in Group 1? ______________
9. Li is the abbreviation for what element? ____________________
10. What element comes after neon, Ne? ____________________
11. Some of the elements have abbreviations that spell short one- or two-letter words. What are they? ____________________

    ____
12. Nickels are made from the element nickel. Give its atomic number and abbreviation. ____________________
13. Calcium is found in milk and is important for bones and teeth. Give its atomic number and abbreviation. ____________________
14. Which period has the fewest elements in it? ____________
15. If carbon has six neutrons and six protons, what is its atomic weight? ________
16. How many electrons does carbon have? ____________
17. What element has the abbreviation Co? (Hint: It's in the 4th period.)

    ____
18. What element is the same number as your age? ______________
19. What element is the same number as your parent or teacher's age?

    ____
20. Find Sulfur, S, on the chart. What element is next to it on the left?

    ____

It's ELEMENTary Lab: Eating Hockey Pucks - instructions

CAUTION: This lab requires use of an oven. Only the parent/instructor should use the oven.

Materials:

- Lab sheet, pencil
- 2 cups All-purpose flour or whole-wheat flour
- ½ teaspoon Salt
- ½ to 1 teaspoon Cinnamon
- Two large eggs
- ⅔ cup Granulated sugar
- 1 cup Milk
- ¼ cup Vegetable oil
- 1 tablespoon Baking powder
- Cinnamon sugar (optional)
- Muffin pan
- Twelve muffin cup liners
- Two mixing bowls
- Whisk
- Measuring cup
- ¼-cup Measuring cup, gravy ladle, or muffin scoop
- Measuring spoon
- Oven with a timer
- Hot pads

Aloud: You would not eat a hockey puck, would you? What would a muffin taste like if it was not fluffy inside? It would probably taste better than a hockey puck, but would it still taste good?

Baking powder is added to baked goods to make them become fluffy. This makes baking powder a leavening agent. A leavening agent is something you add when you are baking to make the finished product, in this case the muffin, fluffy. Baking powder has sodium, hydrogen, carbon, aluminum, sulfur, and oxygen atoms in it. These six different elements are six of the eighteen most commonly occurring elements.

The baking powder changes in the muffin batter into other molecules. One of the molecules it makes is a gas molecule with carbon and oxygen in it, called carbon dioxide (KAR-ben dye-OX-ide). You might remember carbon dioxide if you studied the respiratory system before; plants take it in and people breathe it out. This gas molecule makes baked goods fluffy as they float up. It is the same molecule that makes soda fizz.

What if we made muffins and left out the baking powder? Would the muffins still rise? Could we tell the difference between muffins made with baking powder and muffins made without baking powder? Let's experiment and find out.

Procedure:

1. Complete the hypothesis portion of the lab sheet.
2. Preheat the oven to 375°.
3. Prepare the muffin pan by putting in muffin cup liners.
4. In the first bowl, add 2 cups of flour, ½ teaspoon of salt, and up to 1 teaspoon of cinnamon to the flour, depending on how much you like cinnamon. Whisk the dry ingredients together. (DO NOT ADD THE BAKING POWDER YET.)
5. In the other bowl, add 2 eggs + 1 cup milk + ⅔ cup sugar + ¼ cup vegetable oil. Whisk the wet ingredients together.
6. Stir the dry ingredients into the wet ingredients.

(continued on the back)

7. Use the ¼ cup measuring cup, gravy ladle, or muffin scoop and measure about half of this batter into 6 muffin cup liners.
8. Add 1 tablespoon of baking powder to the remaining batter in the bowl. Stir the batter with the whisk. Do not over mix. Spoon about ¼ cup of batter into each of the remaining 6 muffin liners.
9. Sprinkle the tops of all the muffins with cinnamon sugar, if desired.
10. Put the muffins in the oven for 20 to 25 minutes.
11. Let the muffins cool and cut them in half to observe the inside.
12. Serve each student a leavened muffin and an unleavened muffin (or pieces of each).
13. Record your observations before eating the muffins.

Possible Answers:

Results and Observations:

The muffins looked and tasted different.

Labeled drawings should show that the inside of the leavened muffin looks a bit as if soda was bubbled through it. The unleavened muffin should look smaller and more dense, without bubble holes.

Discussion/Conclusion:

When the muffins with baking powder were cooked, a gas (carbon dioxide) bubbled through them. This gas made these muffins fluffier than the muffins made without carbon dioxide.

Instructor's Notes:

- Cut the leavened muffin in half so you can see the difference between the two types of muffins from the inside.
- The gas molecule that makes baked goods rise is carbon dioxide. Not all baking powder is made from the exact same ingredients. All baking powder does release carbon dioxide, though. Here is one common composition and reaction of baking powder: $NaAl(SO_4)2 + 3\ NaHCO_3 \longrightarrow Al(OH)3 + 2\ Na_2SO_4 + 3\ CO_2$ (The 2 types of molecules on the left side of the equation are the baking powder. The results on the right side are aluminum hydroxide, sodium sulfate, and the important carbon dioxide molecules.)
- This recipe makes about 14 total muffins (7 leavened and 7 unleavened). The recipe can be doubled if you need more muffins.

NAME ________________________________ DATE ________________

It's ELEMENTary Lab: Eating Hockey Pucks

Hypothesis: Do you think there is going to be a noticeable difference between the muffin with baking powder and the muffin without baking powder?

Yes No I don't know

Results and Observations: How are the muffins different?

Taste: ________________________________

____

____

Smell: ________________________________

____

____

Do they look different? Draw labeled pictures of the two different muffins.

Discussion and Conclusion: Describe what happened to the two different muffins. Why did baking powder make a difference? Which muffin did you prefer? Do you think people should keep using baking powder when they make muffins?

____

____

____

____

The Chemist's Alphabet Defined - Crossword Vocabulary Review

Across

4. The rows of the periodic table.
5. The columns of the periodic table, also called chemical families.
6. The number of protons in the nucleus of an atom.
7. The abbreviation of an element's name.
8. He invented the periodic table.

Down

1. The word scientists use for weight.
2. The number of protons plus the number of neutrons in an atom. (Two words)
3. Chart of the elements. (Two words)

Unit 4

The Chemist's Alphabet Applied

Eighteen Elements Hiding in a Poem

The first eighteen elements want to play a game.
They are hiding in this poem,
Each using his nickname.

He
Likes
Nachos
And she does, too!

Be
"**Mg**nificent"
In everything you do.

Bumblebees
Alight
And pollinate the flowers.

Let's be
Constantly
Silly
And dance in April showers.

Nice
Penguins
Say, "Thanks!"

Obnoxious
Seagulls
Play pranks.

Kids are
Frequently
Clever,
Just like you.

He
Never
Argues,
He has better things to do.

NAME ____________________________ DATE ________________

For my notebook

Element Book

by ________________________

Element Book - instructions

As an ongoing project for this unit, your student will be assembling a book with a cover to house his periodic table and information about several chemical families. This book will be assembled in this lab, and then one new page will be decorated each time your student studies a new group or family on the Periodic Table of the Elements. Your student's periodic table, started in the last unit, will be pasted to the first two-page spread.

Materials:

- Copy of "Element Book" cover page
- "My Periodic Table" (two pages) started in the last unit
- Ten sheets of 12″ x 12″ card stock (different colors)
- Colored pencils
- Scissors
- Glue gel or glue stick
- Stapler

Aloud: As we study different chemical families on the Periodic Table of the Elements, you will be putting together a book to show what you have learned. Today you will decorate the cover, assemble the book, and paste your periodic table in the front of the book.

Procedure:

1. Decide in which order you want to assemble the card stock pages. The front and back covers will remain uncut.
2. You will be cutting off rectangles from the top right of 8 pages to make tabs in the book. Hint: Mark your cuts before cutting—not all 12″ x 12″ paper is exactly that.

Page 1 - no cutting; this is the front cover

Page 2 - measure down 1″ from the top and over 10" from the right. Cut this 10″ x 1″ rectangle off. This will form the first tab, on the left side which should measure 2″ wide. This will be the largest tab; all the other tabs will measure 1 ¼″ wide.

Page 3 - measure down 1″ from the top and over 8 ¾″ from the right. Cut this rectangle off to form the second tab, which will be 1 ¼″ wide when page 3 is placed under page 2.

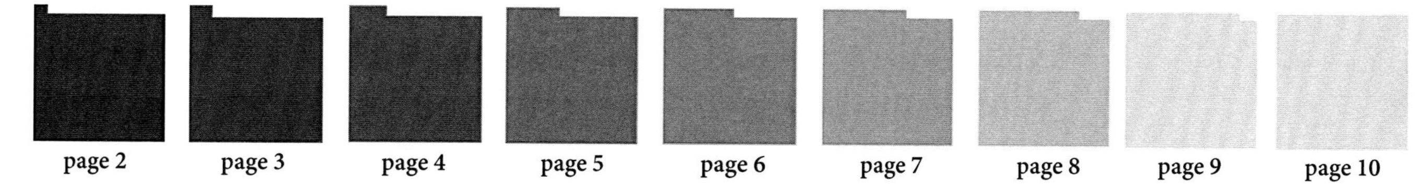

Page 4 - measure down 1″ from the top and over 7 ½″ from the right. Cut this rectangle off to form the third tab, which will be 1 ¼″ wide when page 4 is placed under page 3.

Page 5 - measure down 1″ from the top and over 6 ¼″ from the right. Cut this rectangle off to form the fourth tab, which will be 1 ¼″ wide when page 5 is placed under page 4.

Page 6 - measure down 1″ from the top and over 5″ from the right. Cut this rectangle off to form the fifth tab, which will be 1 ¼″ wide when page 6 is placed under page 5.

Page 7 - measure down 1″ from the top and over 3 ¾″ from the right. Cut this rectangle off to form the sixth tab, which will be 1 ¼″ wide when page 7 is placed under page 6.

Page 8 - measure down 1″ from the top and over 2 ½″ from the right. Cut this rectangle off to form the seventh tab, which will be 1 ¼″ wide when page 8 is placed under page 7.

(continued on the back)

stacked pages

Page 9 - measure down 1″ from the top and over 1 ¼″ from the right. Cut this rectangle off to form the eighth tab, which will be 1 ¼″ wide when page 9 is placed under page 8.

Page 10 - no cutting. When page 10 is placed under page 9, you will see a ninth tab.

3. Assemble the book by stacking the sheets in order so all the tabs line up, the front cover is on the top, and the back cover is on the bottom. Staple in four places about ½″ apart along the left side.
4. Color, decorate, and cut around the "Element Book" cover page. Add your name and glue to the outside front cover. Add trim, drawings, or whatever you have to make your cover irresistible.
5. If you haven't already, remove the two pages of "My Periodic Table" from this book. Trim the pages along the edges as indicated on each page. Glue the pages inside the "Element Book" on the first two page-spread, lining up the two pages along the inside spine (leave a little gap so the book will close easily).
6. On the first tab in your Element Book, cut out and glue the label "My Periodic Table of the Elements" found on page 149.

For More Lab Fun:
Take pictures along the way, doing a lab from each section. Add the pictures to the appropriate page in your "Element Book."

NAME ____________________ DATE ____________

For my notebook

He Likes Nachos - Group 1

The first chemical family you are going to meet is called Group 1, and it has seven elements in it. You are going to meet the three top elements in this group—Hydrogen (H), Lithium (Li), and Sodium (Na), or you could call them **H**e **Li**kes **Na**chos! Let me introduce you.

The first element from Group 1 is Hydrogen (HI-dreh-jen).

Nickname: H

Best Known For: Well, H is #1 on the periodic table. You know what they say: "There can only be one #1."

H is the smallest atom on the periodic table, too. Some of you might be the smallest in your family, but are you the smallest of everybody you know?! H likes it. He is sort of famous, after all. You see, he hangs out with his best friend oxygen, (whose nickname is O) H + H + O = H_2O, which makes the molecule water. How is that for a winning combination?

H is the most common element in the universe.

H is the 10th most common element in Earth's crust.

WOW! It sounds like H is pretty cool. If you are the smallest, just remember, as little H likes to say, "Size is not everything. Or maybe it is, and smallest is best."

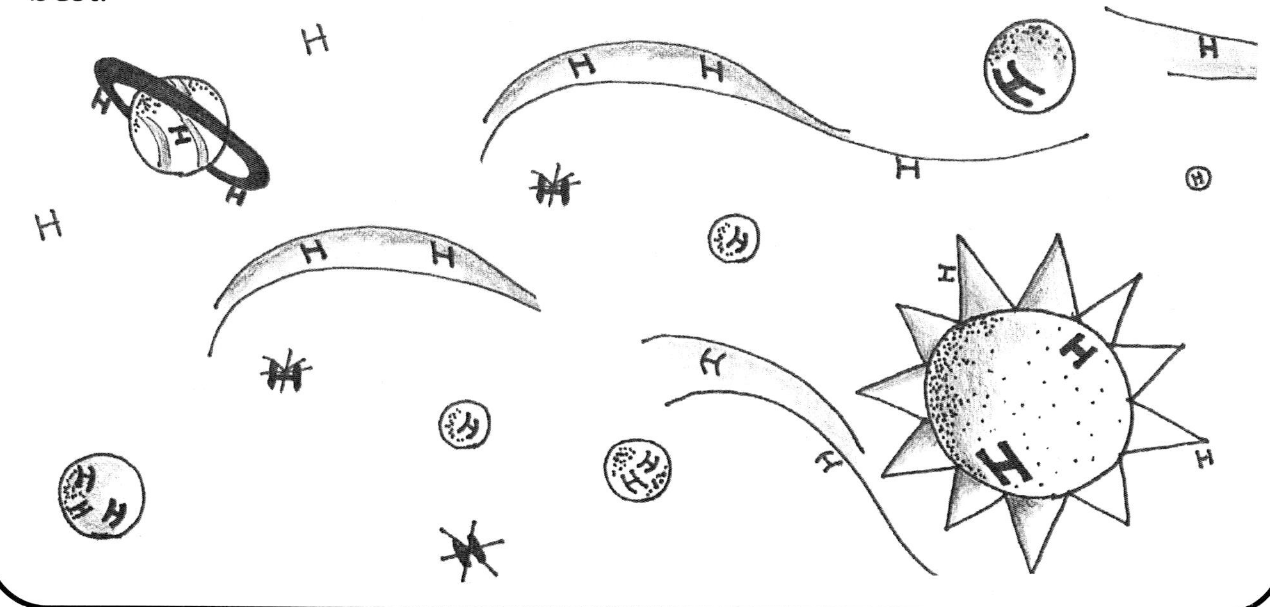

The next element from Group 1 is Lithium (LITH-i-em).
Nickname: Li

Best Known For: Li is named after the Greek word for stone: "lithos." Do you remember the lithosphere when you studied space?

Unlike H, Li is not very common, with only very small amounts found on Earth.

Li is used to make medicines, ceramics, and even batteries.

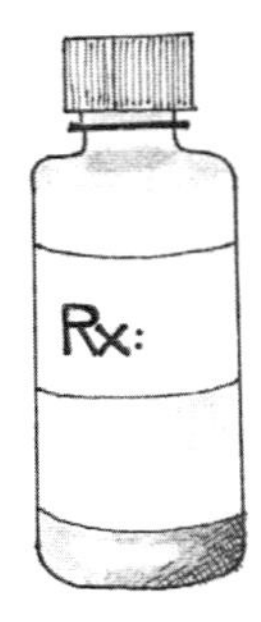

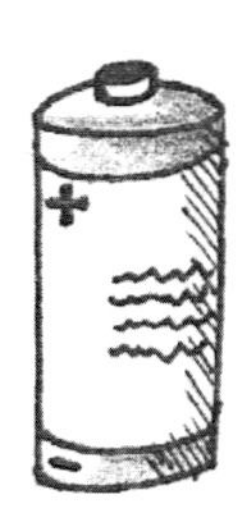

The next element from Group 1 is Sodium (SO-di-em).
Nickname: Na

Best Known For: Na spends a lot of time hanging out with his best friend Chlorine (whose nickname is Cl), Na + Cl = NaCl (sodium chloride), which is table salt. Without NaCl, potato chips wouldn't taste good and the ocean would not be salty.

Na is the sixth most common element in Earth's crust.

Na is found in baking soda and baking powder, and it helps make your biscuits rise.

He Likes Nachos - Group 1 Lab:
The Incredible Floating Egg - instructions

Materials:
- Lab sheet, pencil
- Two raw eggs (with no cracks)
- Two identical tall clear glasses
- Bottle of distilled water (about 2 cups)
- Table salt (about half a cup)
- Teaspoon
- Stirrer (a chopstick works well)

Aloud: When you salt your french fries, you are putting NaCl, sodium chloride, on them. Yum! Let's see what else sodium is good for. Do you think you can make an egg float by adding salt to water? Have you ever heard someone talk about floating better in the ocean than in a swimming pool? Let's find out if things float better in salt water than in water without salt.

Procedure:
1. Complete the Hypothesis portion of the lab sheet.
2. Pour one cup of distilled water into each glass.
3. Put an egg in one of the glasses as a reference control to show how an egg behaves in fresh water. Explain to students that a control model is often used in lab experiments for comparison purposes.
4. In the other glass of water, add salt in 1-teaspoon increments. Stir the salt-water mix to dissolve the salt after each increment, and test the egg to see if it will float. Continue to add salt in 1-teaspoon increments until it will float. Keep track of how many teaspoons of salt it took to float the egg, and record this number on the lab sheet.
5. Draw a picture of each glass with the egg in it. Don't forget to label your pictures.

For More Lab Fun:
When the egg floats, you can add a little water and watch it sink again. When it sinks, you can a little more salt and watch it float. If you are careful about the amount of water you added, you can lightly sprinkle salt on the egg after it has sunk and watch it "dance" as you make it float again.

Instructor's Notes:
- When salt is added to water, it increases the density of the solution. The more salt that is added, the more dense the solution. This increase in density makes things float more easily.
- If the egg has any cracks, it will not float.
- How much salt it takes to float an egg will vary due to the type of egg you are using, whether fresh or store-bought (farm-fresh eggs have less airspace inside the egg, tending to be more dense and therefore require more salt), the size of the egg, and the temperature of the water used.

Answers:
Discussion/Conclusion:

1. Sink
2. More
3. Float

NAME ______________________________ DATE ________________

He Likes Nachos - Group 1 Lab: The Incredible Floating Egg

Hypothesis:

I think an egg will float better in (circle one)

Salted water Fresh water

Results/Observations:
It took ______ teaspoons of salt dissolved in water to float my egg.

Drawings and Labels of the Floating Egg and the Control Egg

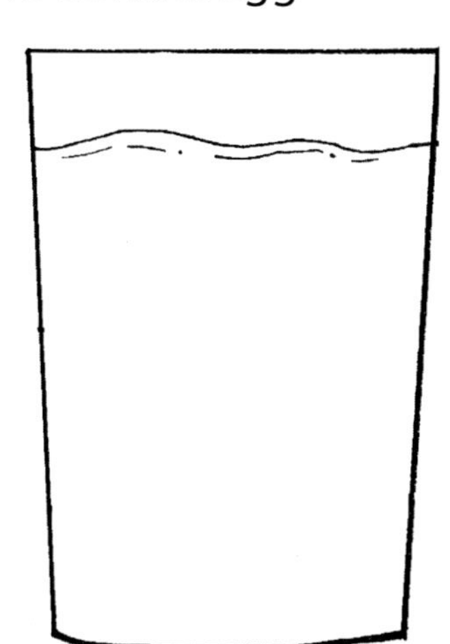

Discussion and Conclusion:
Use the results from this experiment to help answer the following questions (circle the word that makes the sentence correct):

1. If a rock will sink in the ocean, it would (float) (sink) in a freshwater swimming pool.
2. The more <u>buoyant</u> something is, the better it floats. Therefore, eggs in salt water are (more) (less) buoyant than in fresh water.
3. As salt dissolves in water, it helps things in the water to (float) (sink).

Element Book: He Likes Nachos - Group 1 - instructions

Materials:

- Lab sheets (two pages), pencil
- "Element Book" (assembled) with "My Periodic Table"
- Scissors
- Glue
- Art supplies - markers, colored pencils, crayons, etc.
- Flipbook from Unit 3, for reference

Procedure:

1. Cut out the three element symbol squares on the lab sheets. Read the following as you add the three elements to your periodic table.

Aloud: Open up your "Element Book" to your periodic table. Find the number 1 that you wrote above the top square of the first column. The three elements you just learned about are in Group 1 on the periodic table. Remember that all the members of a chemical family (or group) have the same number of electrons in their outer energy levels.

First, color the symbols on the lab sheets for all three elements you met in Group 1.

Next, paste the squares on your periodic table. They are very picky about where they go. The order they go in from top to bottom is H → Li → Na.

Now write their atomic numbers on the squares. Remember, the atomic number tells you the number of protons an atom has. The atomic number of an element goes right above the symbol in its square. H is number 1, as you already know. Write the number 1 above H.

What about the other two Group 1 elements? Do you remember how you count across the periodic table? It's like reading a book. So what number would Li be? Well, if H, hydrogen, is 1, then all the way across the table on the right side would be 2, then Li would be 3. Write the number 3 above Li in its square. Can you figure out what Na would be? (11)

The next step is to draw the electrons in the outer energy level around the elemental symbol. All Group 1 elements have one electron in their outer energy level. That is why they are grouped together. On your periodic table, draw one electron next to each symbol. Hydrogen should look like this: H•. Do this for all of the three elements. Remember that this dot does not tell you the total number of electrons each element has, just that each element in Group 1 has one electron in its outer energy level.

Finally, write the elements' atomic masses. On your periodic table, the atomic mass of an element goes right below the symbol and its name in its square. Remember, the atomic mass of an element is the number of protons (atomic number) an atom has + the number of neutrons. I said that I would always tell you the number of neutrons, but see if you can figure out the atomic mass for each of the three elements in Group 1.

Hydrogen has 0 neutrons
Lithium has 4 neutrons
Sodium has 12 neutrons

Instructor's Notes:

- Create an atomic mass table (like the one found on the back of this page) if it will assist your student in calculating the atomic masses.
- Encourage students to use lots of color on their periodic tables. I chose a different color for each part—green to color the symbol, blue for the atomic number, black for the electrons, and red for the atomic mass. It is best to choose a dark color that really stands out for the electrons.

(continued on the back)

Procedure continued: Decorate Group 1 page in Element Book:

2. Cut out the rest of the items on the lab pages.
3. Use the information you learned this week about the elements to fill in the trivia boxes for each element, listing characteristics, examples, and interesting facts. Color the items as you wish.
4. Create an atom for each element by drawing in the electrons, protons, and neutrons. Draw each part of the atom a different color. Remember that only two electrons go in the first energy level and up to eight electrons in the second and third energy levels. All three elements should only have one electron in their outer energy level. (Refer to your flipbook for help drawing the atoms.)
5. On page 3 of your "Element Book" (the next page after "My Periodic Table"), glue the Group 1 label on the tab. Decorate the page as you like, spreading out the element labels on the page and gluing the other items under their appropriate label.

Answers:

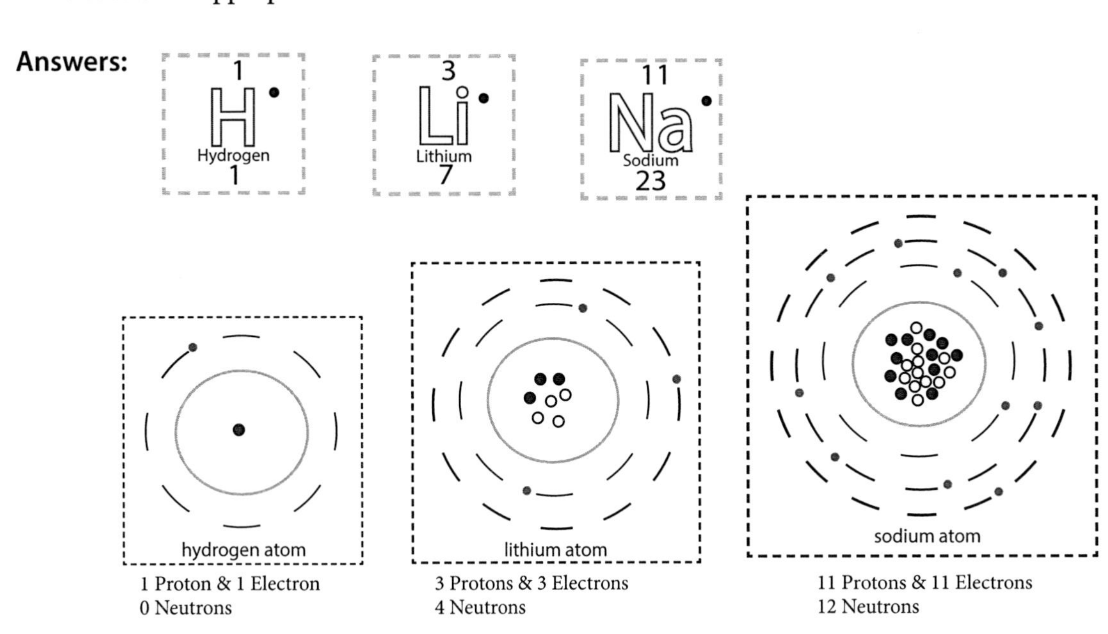

1 Proton & 1 Electron
0 Neutrons

3 Protons & 3 Electrons
4 Neutrons

11 Protons & 11 Electrons
12 Neutrons

● = proton
o = neutron
• = electron

Reminder:
1st energy level can have up to 2 electrons
2nd energy level can have up to 8 electrons

Atomic mass table

Element's Symbol	Atomic Number = Number of Protons	+	Number of Neutrons	=	Atomic Mass
H	1	+	0	=	1
Li	3	+	4	=	7
Na	11	+	12	=	23

Facts on hydrogen: smallest, most common element, winning combination with oxygen to make water.

Lithium lowdown: comes from the Greek word for "stone"; very little found on Earth; used to make medicines, ceramics, and batteries

The truth about sodium: used to make salt, sixth most common in Earth's crust, found in baking soda

NAME ________________________________ DATE ________________

Element Book: He Likes Nachos - Group 1 - page 1

Hydrogen

My Periodic Table of the Elements

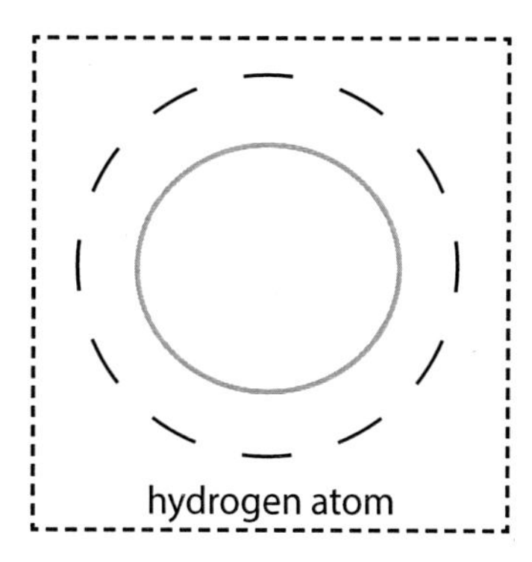
hydrogen atom

Group 1

The Facts on Hydrogen

____

____

____

____

____

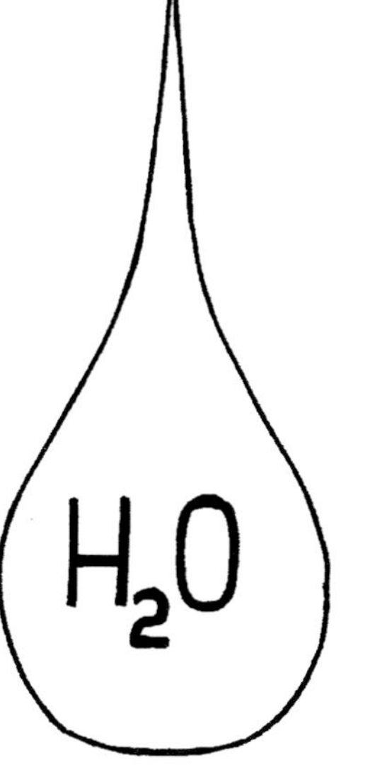

NAME ______________________ DATE ______________

Element Book: He Likes Nachos - Group 1 - page 2

Lithium

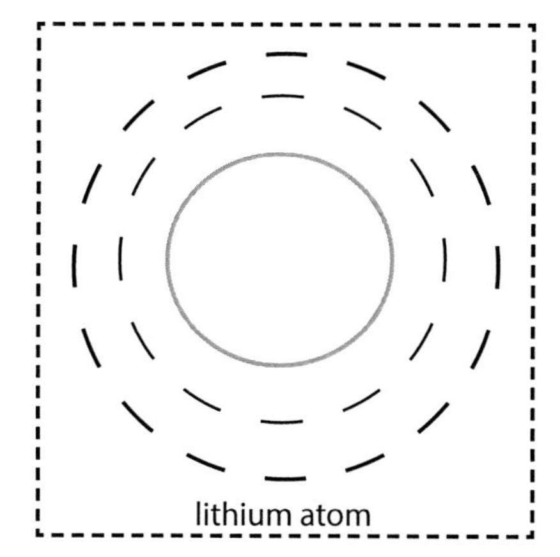

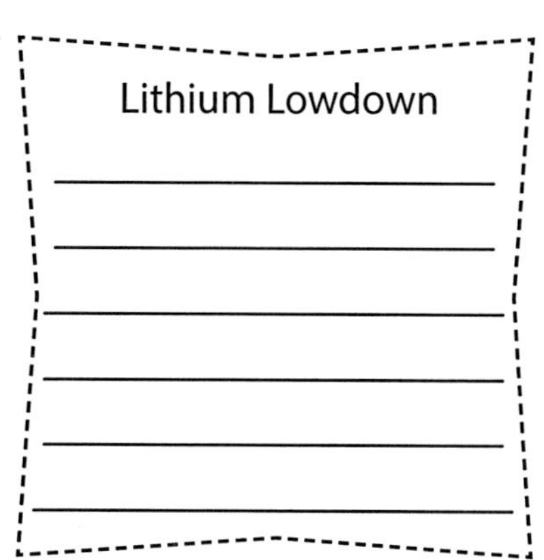

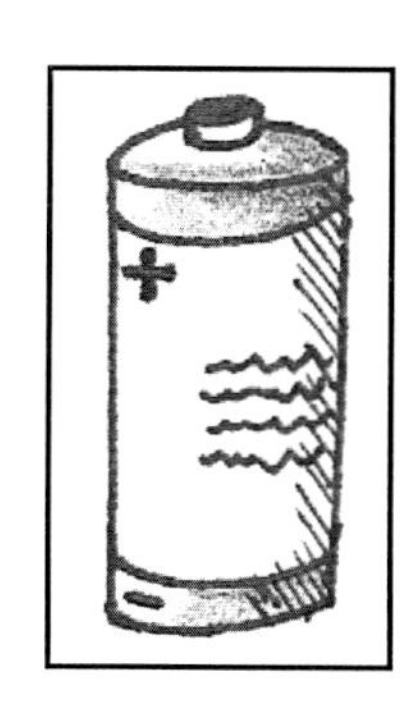

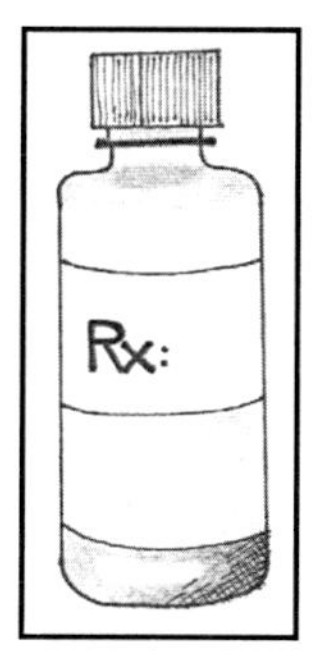

Sodium

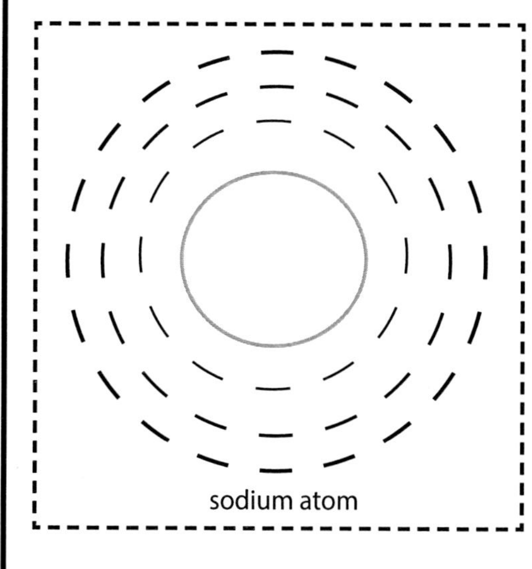

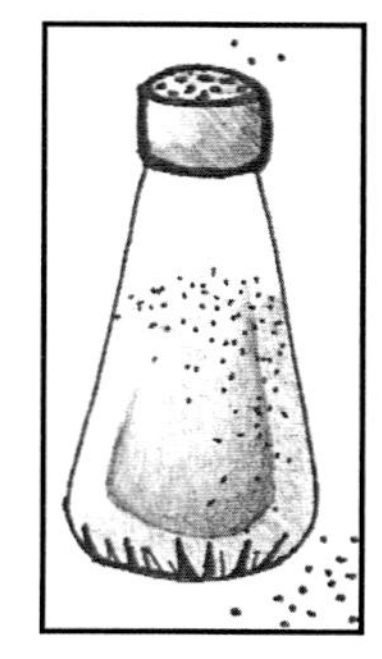

NAME ______________________________ DATE ________________

For my notebook

Be Mgnificent - Group 2

You got along so well with the first chemical family, Group 1, that the second chemical family, Group 2, heard about it and wants to meet you too. There are six elements in Group 2. You are going to learn about two of them: Beryllium (Be) and Magnesium (Mg) or **Be** "**Mg**nificent"

The first element in Group 2 is Beryllium (beh-RIL-ee-em).
Symbol: Be
Best Known For: Be tastes sweet, but Be is a deadly poison if eaten. Be is also one of the elements always found in the green gemstone emerald.

So let me get this straight: If I grind up an emerald and eat it, it will taste sweet, but it can kill me? That's not a good idea!

The next element in Group 2 is Magnesium (mag-NEE-zee-em).
Symbol: Mg
Best Known For: Each cubic mile of seawater has 12 billion pounds of Mg in it. Even if a mile seems like a long way, 12 billion is mega-big!!!

Mg is the eighth most common element in the universe, even if you have never heard of it.

Mg is the seventh most common element in Earth's crust. This makes sense if you think about how much seawater there is.

Mg was discovered in Epsom, England in 1618 by a farmer, when his cows refused to drink water from a well. The farmer tasted the water and found it to be very bitter. He understood why the cows wouldn't drink it. The farmer also found that the water from this well helped cuts and rashes heal. He must have said, "This stuff tastes yucky. I wonder what will happen if I rub it on my cuts?" I don't think his parents were around when he did that, do you? It turns out the water had Epsom salt in it. Epsom salt has Mg in it.

Be Mgnificent - Group 2 Lab: Crystal Creation - instructions

This is a two-day experiment.

Materials:

- Lab sheet, pencil
- 4 tablespoons Epsom salt
- Black construction paper (1 sheet)
- ¼ Warm water
- Magnifying glass or microscope (if you have one)
- Measuring cup
- Tablespoon
- Small cake pan
- Scissors

Aloud: Epsom salt has magnesium in it. Look at the Epsom salt before it is put in water. Do you think the cows can drink it? The Epsom salt needs to dissolve in water first, doesn't it?

When you dissolve Epsom salt in water, you are making a <u>solution</u>. A solution has two parts to it, the <u>solvent</u> and the <u>solute</u>. The Epsom salt is the solute, and water is the solvent because it dissolves the solute. Water is called the <u>universal solvent</u> because it is what most things are dissolved in. In fact, you have a solution running through your body. It is your blood. Water is the solvent in that, too. There are lots of different things, solutes, dissolved in the water in your blood. That is why it is a solution.

When you dissolve Epsom salt in water, let it dry in a sunny place and you will have a crystal creation. When you make crystals by pouring them in a pan and letting them dry, as in this experiment, they are called sheet crystals. This experiment takes two days. The first day you will make the crystal solution. It will not be dry until the next day. Look at the Epsom salt. Do you think if we dissolve it and let it dry in the sun, it will look different? Let's experiment and find out.

Procedure:

Day one:

1. Cut black construction paper to fit the bottom of the cake pan.
2. Sprinkle a 1 tablespoon sample of Epsom salt into a container and set it aside for reference later.
3. Put 3 tablespoons of Epsom salt into ¼ cup of warm water. Stir this until all the Epsom salt dissolves.
4. When all or most of the Epsom salt has dissolved, pour the liquid into the pan. Do not pour any undissolved Epsom salt into the pan.
5. Put this in a sunny place and wait for it to dry.

Day two:

6. When it is dry, compare the Epsom salt crystals you made with the undissolved Epsom salt you set aside. Do they look different?
7. Look at both samples with the magnifying glass. Do they look different?
8. Record your observations on the lab sheet.

Possible Answers:

Yes, the Epsom salt looks different after it is dissolved.
Instead of looking like a salt, it looks like etched glass. Answers in this part will vary.

NAME ______________________________ DATE ______________

Be Mgnificent - Group 2 Lab: Crystal Creation

Yes or No   The Epsom salt looked different after I dissolved it.

If you answered yes, how did it look different?

Draw and Label a picture of how the crystals looked through the magnifying glass. (If you have a microscope, you could look at the crystals under the microscope, too.)

Element Book: Be Mgnificent - Group 2 - instructions

Materials:
- Lab sheet, pencil
- "Element Book" (assembled) with student's periodic table
- Scissors
- Glue
- Art supplies - markers, colored pencils, crayons
- Flipbook from Unit 3, for reference

Aloud: Just like all the chemical families, the elements in Group 2 have properties in common. As with the other families, this is because Group 2 elements have the same number of electrons in their outer energy levels. You can figure out the number of electrons in an element's outer energy level by using the periodic table. Remember from Group 1 that Lithium (Li) has one electron in its outer energy level. Now, put your finger on Beryllium (Be). The atomic number of Lithium is 3 and Beryllium is 4, so you know one proton was added. Since the number of protons match the number of electrons, one electron was also added. 1 electron in the outer energy level + 1 = 2. All the elements in Group 2 have two electrons in their outer energy levels.

Procedure:
1. On the lab sheet, draw the electrons in the outer energy level around the elemental symbols. Remember all Group 2 elements have two electrons in their outer energy levels. That is why they are grouped together.
2. Write the atomic number above each symbol.
3. Write the elements' atomic masses. Create an atomic mass table (like the one found on the back of this page) if it will assist your student in calculating the atomic masses. The atomic mass of an element goes right below the symbol and its name in its square. Remember, the atomic mass of an element is the number of protons (atomic number) an atom has + the number of neutrons. Beryllium has 5 neutrons
Magnesium has 12 neutrons
4. Glue the symbol squares on their appropriate places on the student's periodic table.

Instructor's Note:
- Your student may have figured out by this point that the number of neutrons of an element on the periodic table can be calculated by subtracting the atomic number from the atomic mass. If he has, then congratulate your keen observer. If not, do not worry about teaching this now as it will be taught in RSO Chemistry (level two).

Procedure continued: Decorate Group 2 page in Element Book:

5. Cut out the rest of the items on the lab page.
6. Use information you learned this week about the elements to fill in the trivia boxes for each element, listing characteristics, examples, and interesting facts. Color the items as you wish.
7. Create an atom for each element by drawing in the electrons, protons, and neutrons. Draw each part of the atom a different color. Remember that only two electrons go in the first energy level and up to eight electrons in the second and third energy levels. Both elements should have two electrons in their outer energy levels. (Refer to your flipbook for help drawing the atoms.)
8. On page 4 of your "Element Book," glue the Group 2 label on the tab. Spread out the element labels on the page and glue the other items under their appropriate label.

(continued on the back)

Answers:

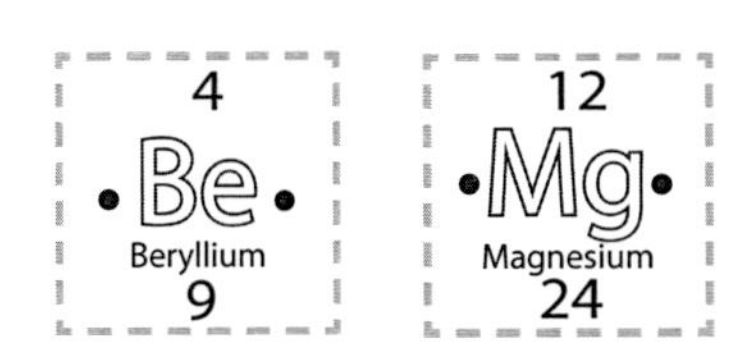

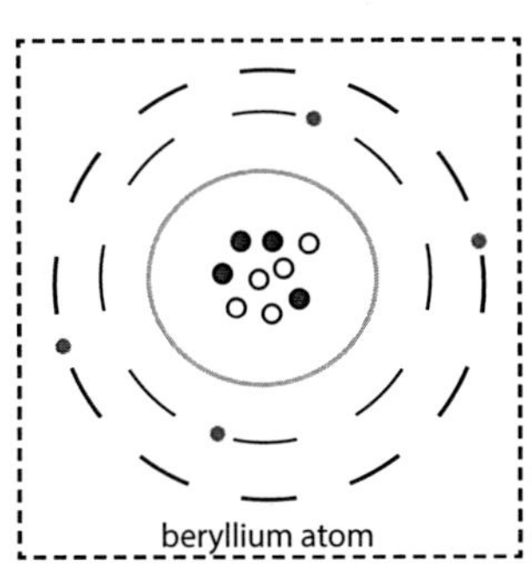

4 Protons & 4 Electrons
5 Neutrons

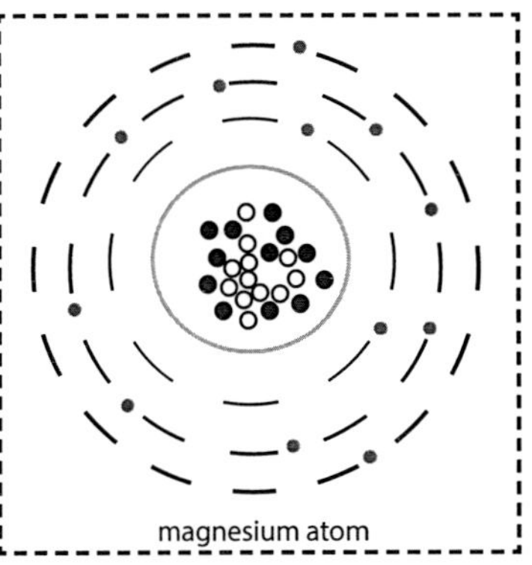

12 Protons & 12 Electrons
12 Neutrons

● = protons
○ = neutrons
• = electrons

Atomic mass table

Element's Symbol	Atomic Number = Number of Protons	+	Number of Neutrons	=	Atomic Mass
Be	4	+	5	=	9
Mg	12	+	12	=	24

Beryllium blow by blow: tastes sweet, deadly poison, found in emeralds

Magnificent magnesium: each cubic mile of seawater has 12 billion pounds of Mg in it, eighth most common element in the universe, seventh most common element in Earth's crust, discovered in 1618 by a farmer, has healing properties.

NAME ______________________ DATE ______________

Element Book: Be Mgnificent - Group 2

Beryllium

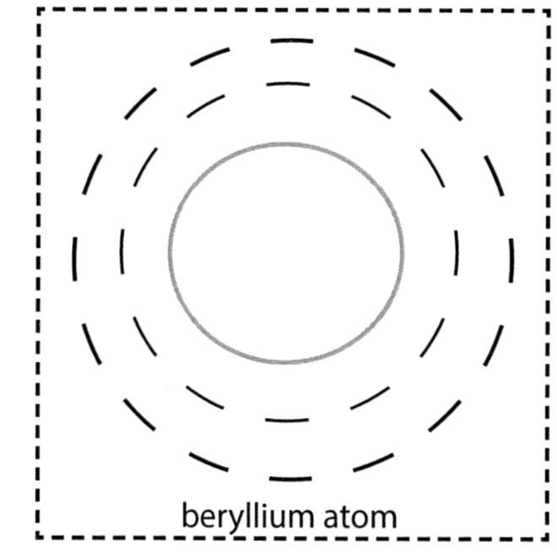

Beryllium Blow by Blow

____

____

____

____

Group 2

Magnesium

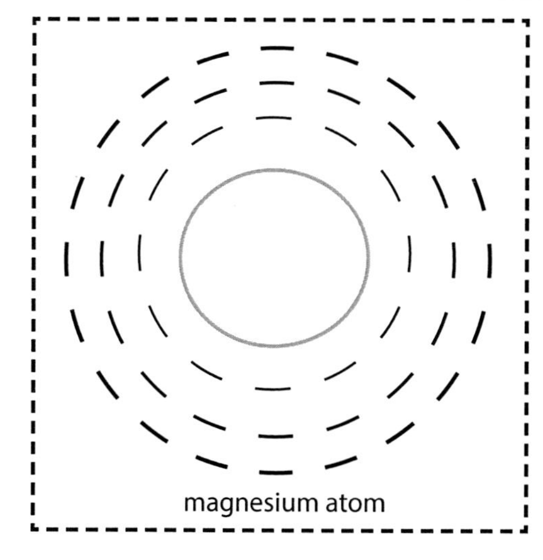

Magnificent Magnesium

____

____

____

____

NAME ______________________________ DATE ________________

For my notebook

Bumblebees Alight - Group 13

Boron and Aluminum from Group 13, or **B**umblebees **Al**ight, want to meet you next, but they can be a little hard to find. You can find Group 13 if you run your finger from beryllium across a space on the periodic table to the next empty square. You can also count all the columns to the 13th column.

The first element in Group 13 is Boron (BO-ron).

Symbol: B

Best Known For: B has a famous line in a Shakespeare play: "To **B** or not to **B**: That is the question."

Marco Polo (1254–1324) brought borax (a compound with B in it) to Europe from the Far East along with gunpowder and spaghetti (which don't have B in them).

There is a **B**ig space between boron and beryllium, but they are still neighbors nonetheless.

B is used in fireworks to make a green color.

The second element in Group 13 is Aluminum (ah-LOO-min-um).

Symbol: Al

Best Known For: Al is the most common metal in the surface of Earth! That must be why they chose Al to make all those soda cans.

Al is the third most common element in Earth's crust.

Al is strong, light, and corrosion-resistant. That means it doesn't rust and change like some metals do. Because of these properties, rockets and large airplanes are made from it.

Bumblebees Alight - Group 13 Lab: The Slime That Ate Slovenia - instructions

Materials:

- 2 teaspoons 20 Mule Team Borax (found next to laundry detergent in stores)
- ¾ cup Water
- ¼ cup White school glue (Elmer's glue works best)
- 1 cup grated Styrofoam or Polystyrene beads*
- Food color (optional to make colorful putty; if you use food color, it can stain surfaces)
- One gallon-size sealable baggie
- Small mixing bowl
- Measuring cup
- Teaspoon
- Tablespoon
- Fork for stirring the glue + water mixture

There is no lab sheet for this lab.

Aloud: 20 Mule Team Borax has boron in it. The borax reacts with the glue to form slime. Pretend you can feel the boron atoms in the slime. Do you think Marco Polo knew about slime?

Procedure:

1. Pour ½ cup of water into the mixing bowl.
2. Put 2 teaspoons of the 20 Mule Team Borax into the water. Stir to help dissolve the borax. Let this sit for 5 to 10 minutes to make sure it is completely dissolved.
3. Measure ¼ cup water into the measuring cup.
4. Add ¼ cup glue into the water so that the water + glue mixture is about ½ cup.
5. Stir the water + glue mixture gently so that you don't spill it. If you are using food color, put it into this mix now.
6. Pour the glue/water solution into the baggie and add 2 tablespoons of the borax/water mixture; don't mix it yet.
7. Add the polystyrene beads to the baggie. Squeeze out the air, tightly seal the baggie, and then knead the mixture gently with your hands. The periodic putty will start to change in consistency very quickly.
8. When the polymer has set up (this should happen within 15 minutes), take it out of the bag.
9. Students can mold the slime and make sculptures. They will feel the beads, "boron atoms," in the putty.
10. If left out, your sculpture(s) will dry nicely and can be painted, or the polymer can be stored in the baggie in the refrigerator to be played with later.

Instructor's Notes:

- *Polystyrene beads can be hard to find. They are used as stuffing for sewing projects and can sometimes be found in the sewing section of craft stores. You can make them easily enough (and cheaper) by taking a piece of Styrofoam and grating it with a cheese grater. If you make your beads this way, there is some clean-up. When you grate the Styrofoam, it builds up an electrostatic charge and sticks to everything, including you.
- This is a recipe that makes a substance similar to Floam. It can be formed into any shape, sculpted around items, allowed to dry, and then painted.
- Use Elmer's glue for the slime. Some other types of white glue do not work in this experiment. The glue needs to have polyvinyl acetate in it, which Elmer's does.
- The polyvinyl acetate in the glue reacts with the borax to form a flexible polymer.
- There is no lab sheet for today—instead students should play with the slime and have fun.
- There is a bonus question in Unit 6 that asks about slime. It would be a good idea to save some slime for this unit by keeping a baggie of it in the fridge.

Element Book: Bumblebees Alight - Group 13 - instructions

Materials:

- Lab sheet, pencil
- Periodic table found on the inside cover of this book
- "Element Book" (assembled) with student's periodic table
- Scissors
- Glue
- Art supplies - markers, colored pencils, crayons
- Flipbook from Unit 3, for reference

Aloud: Just like other families, the elements in Group 13 have properties in common. As with all the chemical families, this is because Group 13 elements have the same number of electrons in their outer energy levels. All the elements in Group 13 have three electrons in their outer energy levels. Can you find boron at the top of Group 13? What is its atomic number? (5) What period is it in? (2nd) Can you find the atomic number and period for aluminum? (13, 3rd period)

Procedure:

1. On the lab sheet, draw the electrons in the outer energy level around the elemental symbol. (See example below.) Remember, all Group 13 elements have three electrons in their outer energy levels. That is why they are grouped together.
2. Write the atomic number above each symbol.
3. Write the elements' atomic masses. Create an atomic mass table (like the one found on the back of this page) if it will assist your student in calculating the atomic masses. The atomic mass of an element goes right below the symbol and its name in its square. Remember, the atomic mass of an element is the number of protons (atomic number) an atom has + the number of neutrons. Boron has 6 neutrons
   Aluminum has 14 neutrons
4. Glue the symbol squares on their appropriate places on the student's periodic table.

Procedure continued: Decorate Group 13 page in "Element Book":

5. Cut out the rest of the items on the lab page.
6. Use information you learned this week about the elements to fill in the trivia boxes for each element, listing characteristics, examples, and interesting facts. Color the items as you wish.
7. Create an atom for each element by drawing in the electrons, protons, and neutrons. Draw each part of the atom a different color. Remember that only two electrons go in the first energy level and up to eight electrons in the second and third energy levels. Both elements should have three electrons in their outer energy levels. (Refer to your flipbook for help drawing the atoms.)
8. On page 5 of your "Element Book," glue the Group 13 label on the tab. Spread out the element labels on the page and glue the other items under their appropriate label.

Answers:

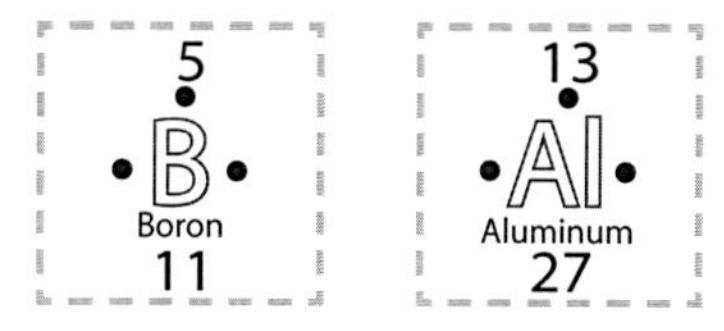

(continued on the back)

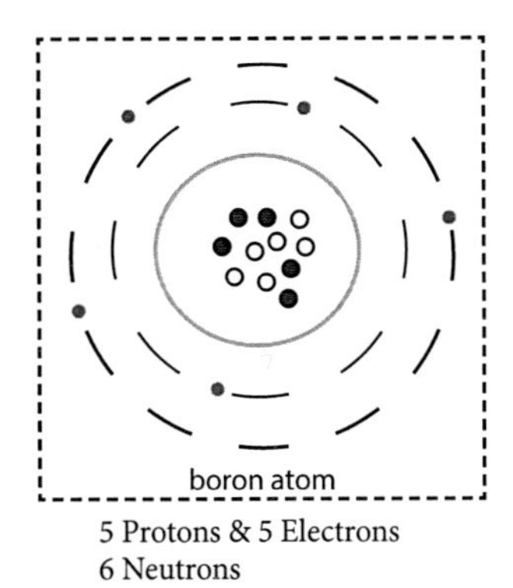

5 Protons & 5 Electrons
6 Neutrons

aluminum atom

13 Protons & 13 Electrons
14 Neutrons

● = protons
○ = neutrons
• = electrons

Atomic mass table

Element's Symbol	Atomic Number = Number of Protons	+	Number of Neutrons	=	Atomic Mass
B	5	+	6	=	11
Al	13	+	14	=	27

Boron basics: Marco Polo brought it to Europe, used to make green fireworks

Aluminum particulars: most common metal in the surface of Earth; 3rd most common element in Earth's crust; strong, light, and corrosion-resistant (doesn't rust); rockets and large airplanes are made from it.

NAME ______________________ DATE ______________

Element Book: Bumblebees Alight - Group 13

Boron

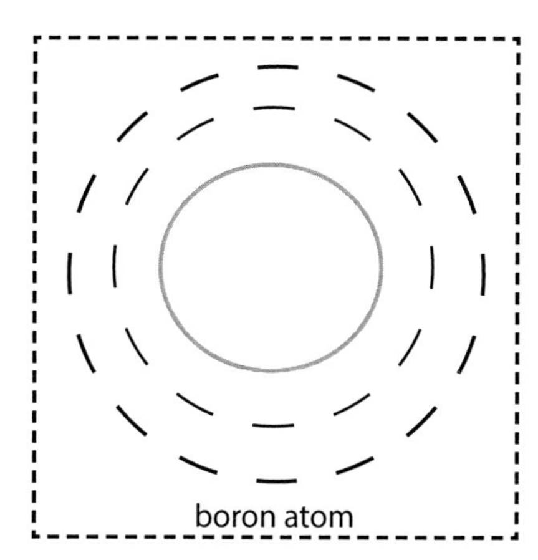
boron atom

Boron Basics

____

____

____

____

Aluminum

Group 13

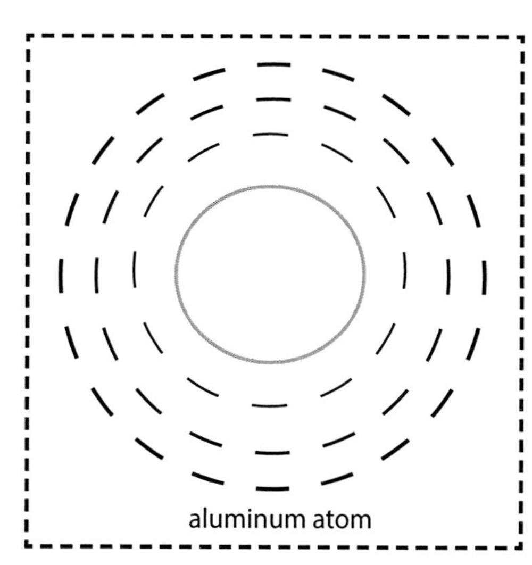
aluminum atom

Aluminum Particulars

____

____

____

____

NAME ______________________________ DATE ________________

For my notebook

Constantly Silly - Group 14

Group 14 has two very important elements in it: carbon and silicon. I like to call them **C**onstantly **Si**lly. A scientific way to talk about living things on Earth is to say, "On Earth we have carbon-based life forms." That means living things use carbon for energy and to grow. Silicon is important because sand is made of silicon + oxygen, and there is a LOT of sand on Earth.

The first element in Group 14 is Carbon (KAR-ben).
Symbol: C

C comes in two elemental forms. Remember that an element has only one type of atom. Pick up a pencil. You are looking at one of the elemental forms of C. The pencil lead (the part of the pencil that writes) is pure C. This form of C has a special name, called graphite. Does your mother or teacher have a diamond ring? If they do, look at the diamond in it. The diamond in the ring is the other elemental form of C. Pencils' leads break easily, but diamonds are the hardest natural substance known. Living things, graphite, and diamonds—C is kind of **C**ool, don't you think?

Best Known For: Living things on Earth all have C in them, from bees to trees to knees. C is the sixth most common element in the universe.

The second element in Group 14 is Silicon (SIL-i-kon).
Symbol: Si

Best Known For: Been to the beach lately? Si is a **Si**gnificant part of any beach. There Si likes to hang out with oxygen (O). O is popular isn't she? You will meet O later. $Si + O + O = SiO_2$, which is sand. Do you know what happens when sand is heated at a high temperature? It melts, and melted sand is used to make glass. Over 3,500 years ago, people were heating sand to make glass-like objects. In addition to sand, quartz, amethysts, flint, opals, and bricks are made from silicon.

Si is the seventh most common element in the universe.

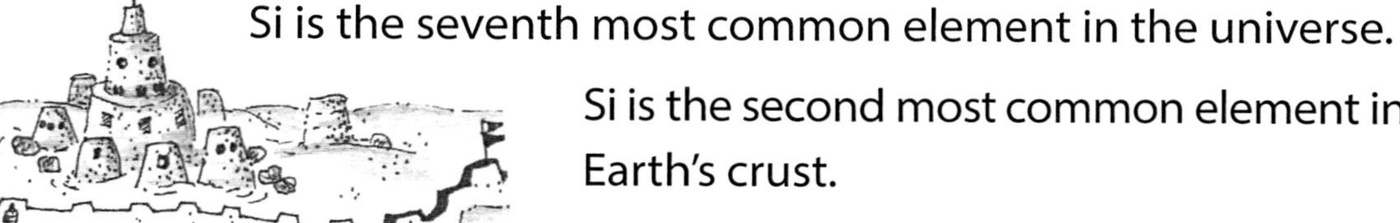

Si is the second most common element in Earth's crust.

Constantly Silly - Group 14 Lab: S'more Carbon - instructions

CAUTION: This lab involves the use of an oven and requires close parental supervision.

Materials:
- Lab sheets (two pages), pencil
- Four large marshmallows
- Two graham crackers
- Baking sheet
- Oven
- Hot pad
- Brown colored pencil

Aloud: When you made marshmallow atoms, did you eat any of the marshmallows? Marshmallows are sweet, aren't they? Do you know what "sweet" means? "Sweet" means sugar, and sugar means carbon! Sugar has 12 carbon atoms, 22 hydrogen atoms, and 11 oxygen atoms in it. When carbon is heated, it starts to darken. At first, it turns a nice golden color. If you heat it too long, though, it turns black.

Have you ever toasted marshmallows? Toasting marshmallows is a good way to see what happens when you heat carbon. That toasty brown color is from the carbon in the marshmallows. I like toasted marshmallows perfectly browned on the outside, and melted and gooey on the inside. How do you like your toasted marshmallows? For this experiment, you are going to brown some marshmallows, and maybe even burn them if that is how you like them.

In addition to sugar, marshmallows have air and water in them. When the air and water in the marshmallows get hot, the marshmallows can grow in size. Have you ever heated water in a pan and watched it steam? When you heat marshmallows, the water in them starts to steam and tries to escape, making the marshmallow expand.

Procedure:
1. Fill in the hypothesis section of your lab sheet.
2. Preheat the oven to 500°.
3. Put a graham cracker on the baking sheet. Put two marshmallows on top of the graham cracker. Make another graham cracker-marshmallow combination, and set this on the counter as a reference.
4. When the oven is preheated, put the baking sheet into the oven on a rack positioned in the middle. Turn the oven light on.
5. Set the timer for 10 minutes. Have your lab sheet ready, you will be making observations every one minute for 8 to 10 minutes.
6. Every one minute, look at the change in color and size. You do not need to take the cookie sheet out of the oven. Just quickly see if you notice any change from the last time you looked.
7. When the marshmallows are browned (or burned if you are going to go that long), put a hot pad on and take the cookie sheet out of the oven.
8. Use your reference graham cracker-marshmallow combination to compare how the marshmallow has changed in size and color.

Instructor's Notes:
- If you have a lower oven with a light, use that for this experiment. Students can sit on the floor while they monitor the color and size changes of the marshmallows.
- The marshmallows are very chewy and hot when they come out of the oven. They do not have the same consistency (and some say they're not as good) as when they are toasted over a fire.
- The marshmallows expand because the air and water in them expand when heated.

(Continued on the back)

Possible Answers:
The marshmallows are going to increase in size, and the color will become increasingly brown.

Discussion/Conclusion:
The marshmallows brown because they have carbon in them, and carbon browns as it cooks. The marshmallows get larger because the water and air in them expand as they get hot.

Challenge Questions:
The graham cracker has sugar and flour in it—both of which have carbon in them. This carbon is turning brown.

Graphite is a type of carbon. Carbon darkens when exposed to heat. (Graphite deposits are found in metamorphic rock. Metamorphic rock has been exposed to temperatures greater than 150° to 200° C—that's 302° to 392° F.)

NAME ______________________ DATE ______________

Constantly Silly - Group 14 Lab: S'more Carbon Lab - page 1

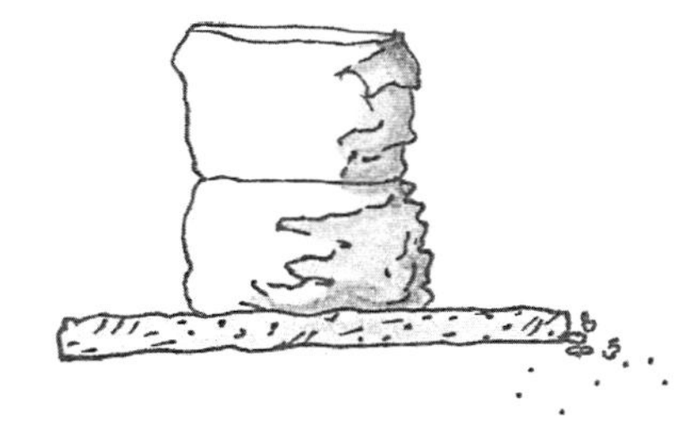

Hypotheses:

My marshmallow at the start of the lab:

How my marshmallow will look at the end of the lab:

Results:

Data Table

Time	Color	Size Change
1 minute		
2 minutes		
3 minutes		
4 minutes		
5 minutes		
6 minutes		
7 minutes		
8 minutes		
9 minutes		
10 minutes		

Key: NC no change
↑ increasing in size or color
↓ decreasing in size or color

Discussion and Conclusion:

How did your marshmallow actually look at the end of this experiment?

Reference marshmallow	Cooked marshmallow

Was this different from your prediction? Yes or No

Challenge Questions:

The graham cracker starts to brown, too. Why?

Graphite (pencil lead) is very dark gray in color. What do you think happened to it to turn it that color?

Element Book: Constantly Silly - Group 14 - instructions

Materials:
- Lab sheet, pencil
- Periodic table found on the inside cover of this book
- "Element Book" (assembled) with student's periodic table
- Scissors
- Glue
- Art supplies - markers, colored pencils, crayons
- Flipbook from Unit 3, for reference

Aloud: Just like other families, the elements in Group 14 have properties in common. As with all the chemical families, this is because Group 14 elements have the same number of electrons in their outer energy levels. All the elements in Group 14 have one more electron in their outer energy levels than those in Group 13 had. Do you remember how many electrons boron and aluminum had in their outer energy levels? (three) So that must mean that Group 14 elements have four electrons in their outer energy levels. Can you find carbon next to boron on the periodic table? It's at the top of Group 14. What is its atomic number? (6) What period is it in? (2nd) Can you find the atomic number and period for silicon? (14, 3rd period)

Procedure:
1. On the lab sheet, draw the electrons in the outer energy level around the elemental symbol. (See example below.) Remember, all Group 14 elements have four electrons in their outer energy levels. That is why they are grouped together.
2. Write the atomic number above each symbol.
3. Write the elements' atomic masses. Create an atomic mass table (like the one found on the back of this page) if it will assist your student in calculating the atomic masses. The atomic mass of an element goes right below the symbol and its name in its square. Remember, the atomic mass of an element is the number of protons (atomic number) an atom has + the number of neutrons. Carbon has 6 neutrons
   Silicon has 14 neutrons
4. Glue the symbol squares on their appropriate places on the student's periodic table.

Procedure continued: Decorate Group 14 page in "Element Book":

5. Cut out the rest of the items on the lab page.
6. Use information you learned this week about the elements to fill in the trivia boxes for each element, listing characteristics, examples, and interesting facts. Color the items as you wish.
7. Create an atom for each element by drawing in the electrons, protons, and neutrons. Draw each part of the atom a different color. Remember that only two electrons go in the first energy level and up to eight electrons in the second and third energy levels. Both elements should have four electrons in their outer energy levels. (Refer to your flipbook for help drawing the atoms.)
8. On page 6 of your "Element Book," glue the Group 14 label on the tab. Spread out the element labels on the page and glue the other items under their appropriate label.

Answers:

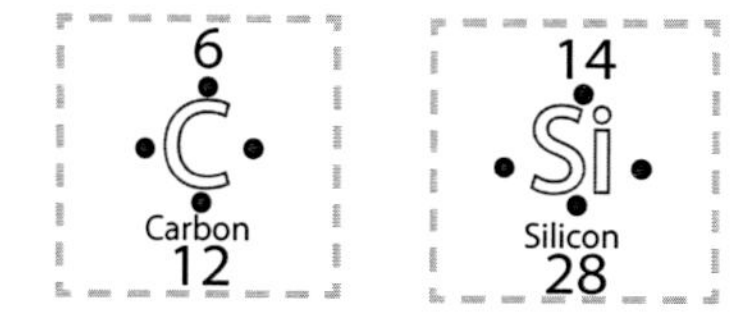

(continued on the back)

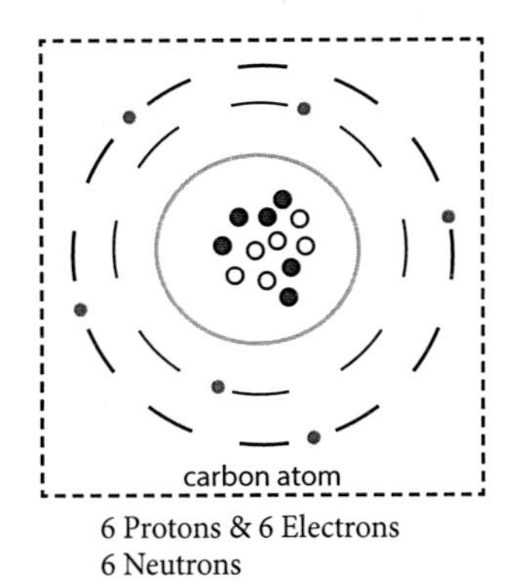

6 Protons & 6 Electrons
6 Neutrons

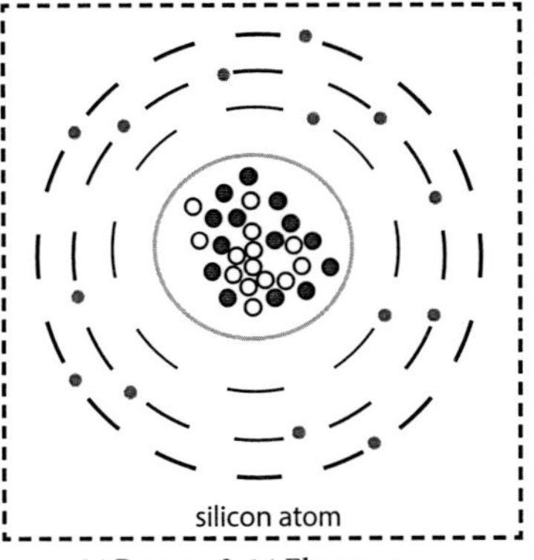

14 Protons & 14 Electrons
14 Neutrons

● = protons
o = neutrons
• = electrons

Atomic mass table

Element's Symbol	Atomic Number = Number of Protons	+	Number of Neutrons	=	Atomic Mass
C	6	+	6	=	12
Si	14	+	14	=	28

Cold hard carbon facts: All life forms have carbon, carbon browns as it is heated up, graphite and diamonds are elemental forms of carbon

Super silicon trivia: Found in sand, quartz, amethysts, flint, opals and bricks; the seventh most common element in the universe; the second most common element in Earth's crust.

NAME ________________________________ DATE ____________________

Element Book: Constantly Silly - Group 14

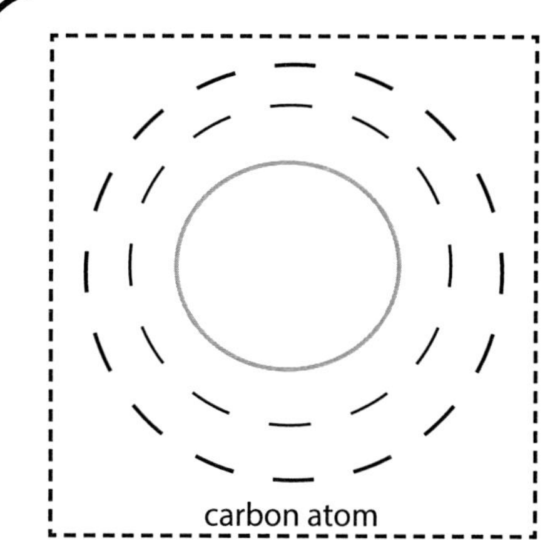

Carbon

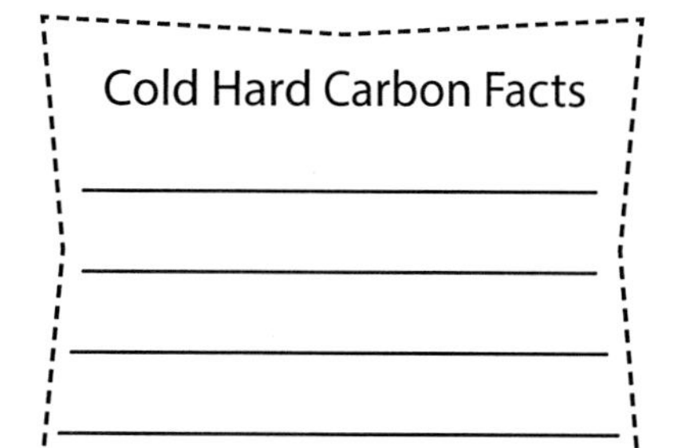

Group 14

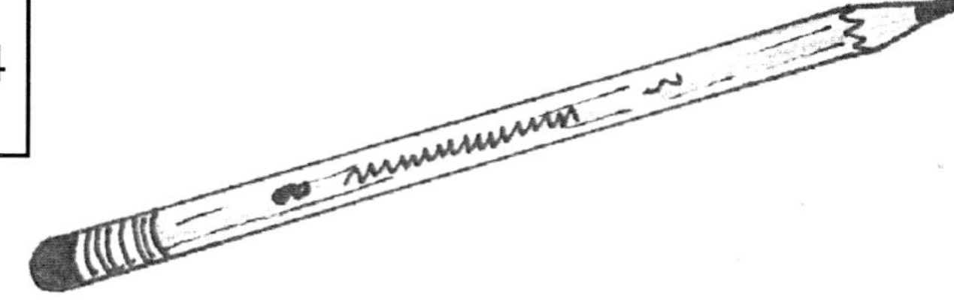

Silicon

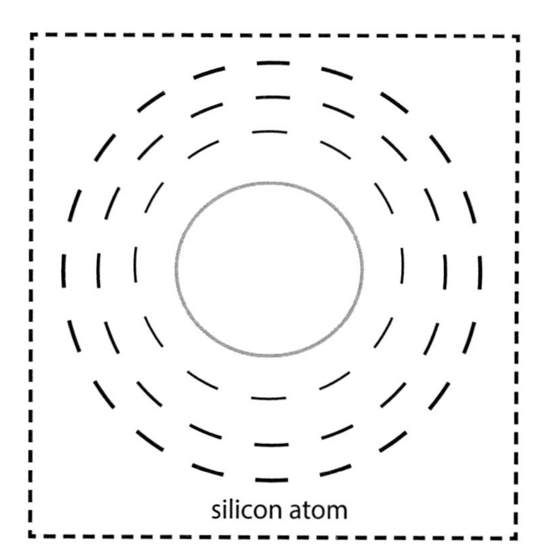

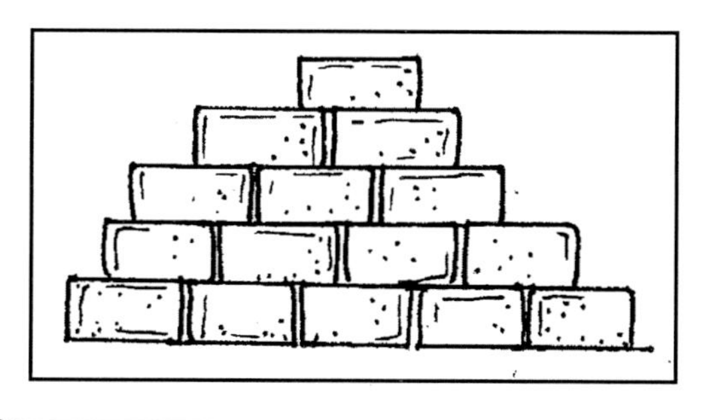

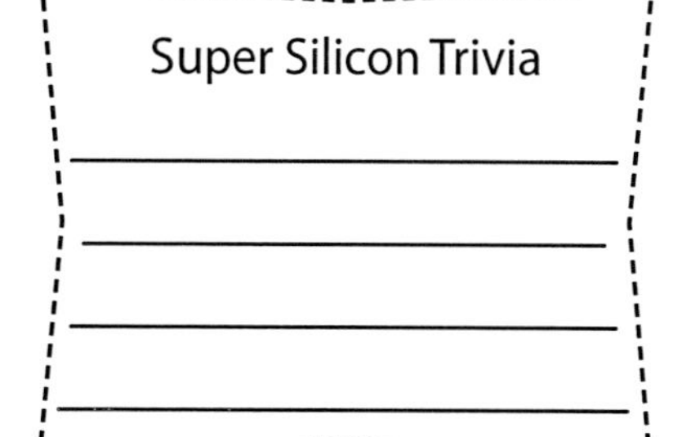

NAME ______________________________ DATE ________________

For my notebook

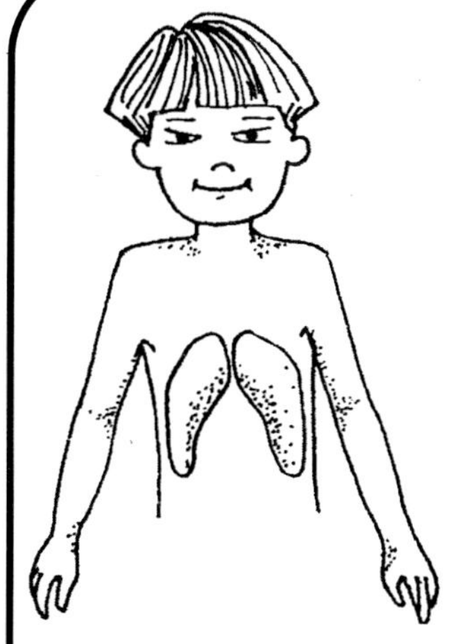

Nice **P**enguins - Group 15

Today you are going to meet two elements from Group 15: nitrogen and phosphorus. Or you could call them **N**ice **P**enguins—aww!

The first element in Group 15 is Nitrogen (NYE-truh-jen).
Symbol: N

Best Known For: Breathe in. Are your lungs full? The air that is in your lungs right now is 78% N. Did you notice the N? Can you feel those little N molecules bouncing around inside of you? Well, that is just what they are doing.

N + H + H + H makes NH_3 which makes ammonia. NH_3 is used for fertilizers and explosives. NH_3 is also used in cleaning products around your house.

Plants need N, and who hasn't heard of TNT? Guess what the N stands for?

N is the fifth most common element in the universe.

N is the most common element in Earth's atmosphere.

The second element in Group 15 is Phosphorus (FOS-fer-iss)
Symbol: P

Best Known For: P is an essential element for all living cells. More important to many of you, though, P is in the explosive part of cap-gun caps.

P has been known since ancient times because rocks with P in them are phosphorescent. That means they glow in the dark! The name phosphorus comes from the Greek word for *light-bearing*.

P is used in fireworks. When P gets hot, it burns really brightly.

Nice Penguins - Group 15 Lab: Eating Air - instructions

CAUTION: This lab involves use of an oven. Only the parent/instructor should operate the oven.

Materials:

- Lab sheet, pencil
- Three egg whites
- ¾ cup Confectionery sugar
- Parchment paper or aluminum foil
- Shortening or nonstick spray
- 1 tablespoon Flour
- Cookie sheet
- Mixing bowl
- Fork
- Spoon
- Electric mixer
- Oven, hot pads
- Magnifying glass

Aloud: Let's eat some nitrogen in the air. Today, you are going to experiment with egg whites and air. First, you are going to examine the white part of a raw egg to see if any air is visible. Then you will whip the egg white. When you do this at a high speed, air is mixed in with the egg whites. With the egg white + air mix, you can make something called meringue. You just have to add a little sugar and you will have a yummy treat to eat.

Procedure:

1. Complete the hypothesis portion of the lab sheet.
2. Crack the eggs and separate the whites from the yolks. You will not need the yolks for this experiment. Put the whites into the mixing bowl. Have students examine the egg whites for the presence of air. They can use a magnifying glass for this. Take a fork and run it slowly through the egg whites. In this part of the experiment, the purpose is to examine the whites for the presence of incorporated air. There should not be any visible pockets of air.
3. Preheat the oven to 225°F. Cover the cookie sheet with parchment paper or aluminum foil. Grease the paper or foil and lightly dust it with flour.
4. Beat the egg whites at the highest speed your mixer will go until they form soft peaks. The amount of time this takes varies depending on the type of mixer you have. Have students examine the egg whites now and record observations on the lab sheet. They can use a magnifying glass.
5. Sprinkle the confectionery sugar over the egg whites and start the mixer at a slow speed until the sugar is mixed in, then increase the speed to maximum. Mix until the egg whites form stiff peaks. Examine the whites again and record observations on the lab sheet.
6. Spoon about 12 spoonfuls of the meringue, as you would cookie dough, onto the prepared cookie sheet. Put the cookie sheet into the oven for 60 minutes or longer, until they feel hard on the outside. When done, turn off the oven and open the oven door, leaving the cookies in the oven, cooling gradually for 5 to 10 minutes.
7. When the meringues are done and have cooled, cut one in half and examine it. Look at all the air trapped inside the egg whites. Record observations on the lab sheet. Meringues can be eaten.

Instructor's Notes:

- Do not taste the egg whites until they have come out of the oven. Until then, they are raw eggs.
- The bowl and the beaters used to mix the egg whites must not have any grease or oil on them.

(continued on the back)

- Meringues can be sensitive to the amount of the humidity in the air. If you live somewhere humid, they can take longer to cook and they will not stay crispy and hard as long as somewhere that has a dry climate.
- The purpose of adding the sugar is to create something edible. The sugar is not essential to the experiment and does not change the outcome.
- As always, have students fill in the results section of this experiment as you perform each step, not at the end of the experiment.

Possible Answers:

Results/Observations:

C1: There will be no visible air in the egg whites when they come out of the shell.

C2: There is air in the egg whites. Whipping the air into them has changed the amount of space the egg whites take up, their texture, and their color. The egg whites are whiter in color and they feel creamy.

C3: The stiffness will be different. However, the egg whites should look about the same as they did at C2.

C4: The egg whites will be hard after cooking. The outsides might have browned (due to the carbon in the sugar). Inside they should look about the same as at C2.

Discussion/Conclusion:

Help students understand that as the egg whites are mixed in air, the air is mixed into the egg whites. The more air that gets mixed in with the egg whites, the more fluffy they become, i.e. the larger the increase in the volume of the eggs. There comes a point when you reach the maximum amount of air that can be mixed in.

Challenge:

In the vacuum of space, there is no air to mix into the egg whites. They would not become fluffy.

NAME ______________________________ DATE ________________

Nice Penguins - Group 15 Lab: Eating Air

Hypothesis: Do you think the egg whites will look different with air mixed into them?

Yes No I don't know

Results/Observations:

Use words and draw pictures to describe the egg whites:

Egg whites before mixing	Egg whites halfway through mixing	Egg whites at the end of mixing	Cooked egg whites

Discussion and Conclusion:

Describe in your own words what happened to the egg whites when you whipped them.

Challenge:

What would happen if you mixed the egg whites in the vacuum of space?

Element Book: Nice Penguins - Group 15 - instructions

Materials:

- Lab sheet, pencil
- Periodic table found on the inside cover of this book
- "Element Book" (assembled) with student's periodic table
- Scissors
- Glue
- Art supplies - markers, colored pencils, crayons
- Flipbook from Unit 3, for reference

Aloud: Using a periodic table, see if you can figure out the atomic number for nitrogen. (7) What is the atomic number for phosphorus? (15) Just like other families, the elements in Group 15 have properties in common because Group 15 elements have the same number of electrons in their outer energy levels. Let's see if you know how many there are. To figure this out, how many electrons are in the outer energy levels of the Group 13 elements? (3) And how many electrons are in the outer energy levels of the Group 14 elements? (4) What is the pattern? Do you know the answer? All the elements in Group 15 have one more electron in their outer energy levels than those in Group 14 had. So that must mean that Group 15 elements have five electrons in their outer energy levels.

Procedure:

1. On the lab sheet, draw the electrons in the outer energy level around the elemental symbol. (See example below.) Remember, all Group 15 elements have five electrons in their outer energy levels. That is why they are grouped together.
2. Write the atomic number above each symbol.
3. Write the elements' atomic masses. Create an atomic mass table (like the one found on the back of this page) if it will assist your student in calculating the atomic masses. The atomic mass of an element goes right below the symbol and its name in its square. Remember, the atomic mass of an element is the number of protons (atomic number) an atom has + the number of neutrons. Nitrogen has 7 neutrons
Phosphorus has 16 neutrons
4. Glue the symbol squares on their appropriate places on the student's periodic table.

Procedure continued: Decorate Group 15 page in Element Book:

5. Cut out the rest of the items on the lab page.
6. Use information you learned this week about the elements to fill in the trivia boxes for each element, listing characteristics, examples, and interesting facts. Color the items as you wish.
7. Create an atom for each element by drawing in the electrons, protons, and neutrons. Draw each part of the atom a different color. Remember that only two electrons go in the first energy level and up to eight electrons in the second and third energy levels. Both elements should have five electrons in their outer energy levels. (Refer to your flipbook for help drawing the atoms.)
8. On page 7 of your "Element Book," glue the Group 15 label on the tab. Spread out the element labels on the page and glue the other items under their appropriate label.

Answers:

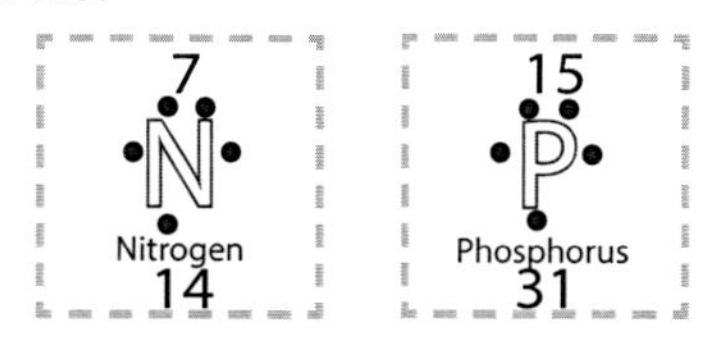

(continued on the back)

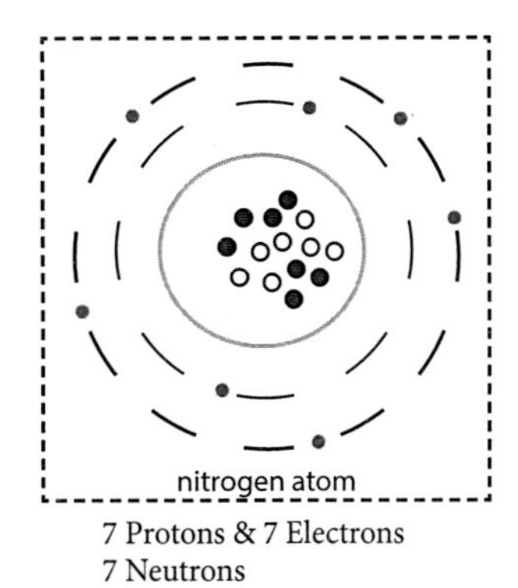

7 Protons & 7 Electrons
7 Neutrons

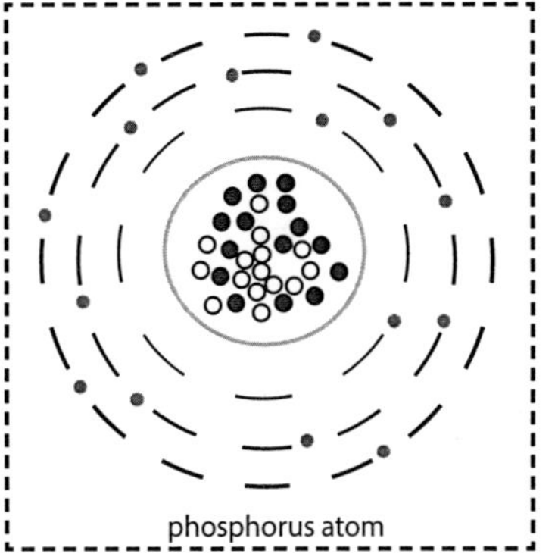

15 Protons & 15 Electrons
16 Neutrons

● = protons
○ = neutrons
• = electrons

Atomic mass table

Element's Symbol	Atomic Number = Number of Protons	+	Number of Neutrons	=	Atomic Mass
N	7	+	7	=	14
P	15	+	16	=	31

The nitty-gritty on nitrogen: 78% of air is N; needed by plants; used to make ammonia, fertilizers, explosives; fifth most common element in the universe; the most common element in Earth's atmosphere

Phosphorus phenomenon: Essential element for all living cells, in the explosive part of cap-gun caps, makes rocks phosphorescent, comes from the Greek word for "light-bearing," used in fireworks.

NAME ______________________ DATE ______________

Element Book: Nice Penguins - Group 15

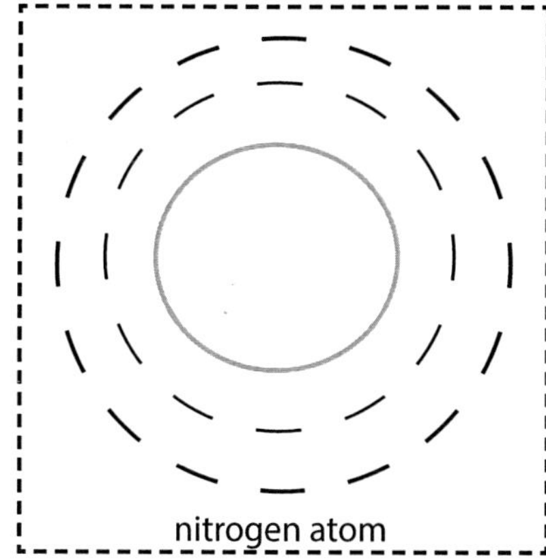

nitrogen atom

Nitrogen

The Nitty-gritty on Nitrogen

____

____

____

____

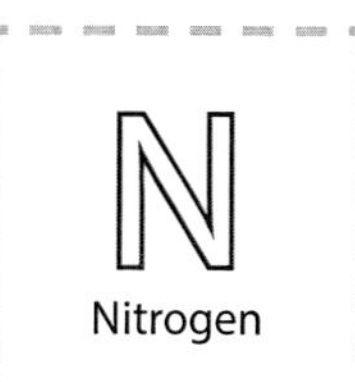

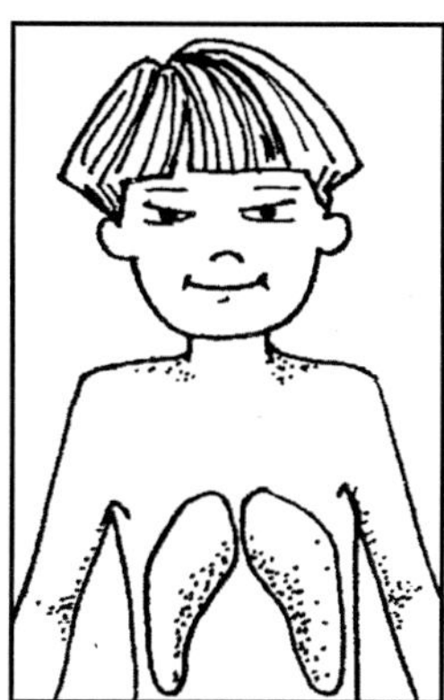

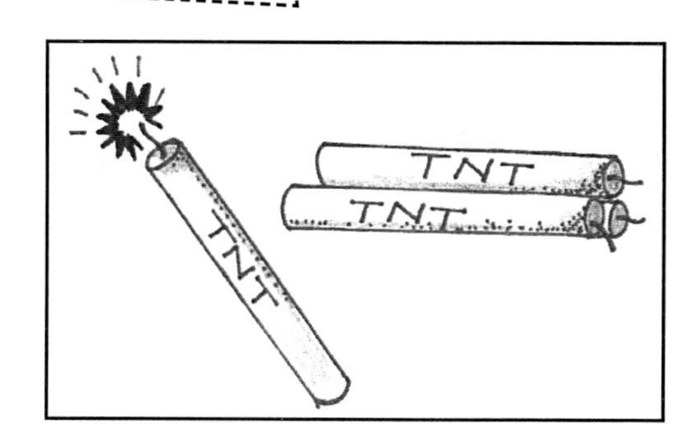

Phosphorus

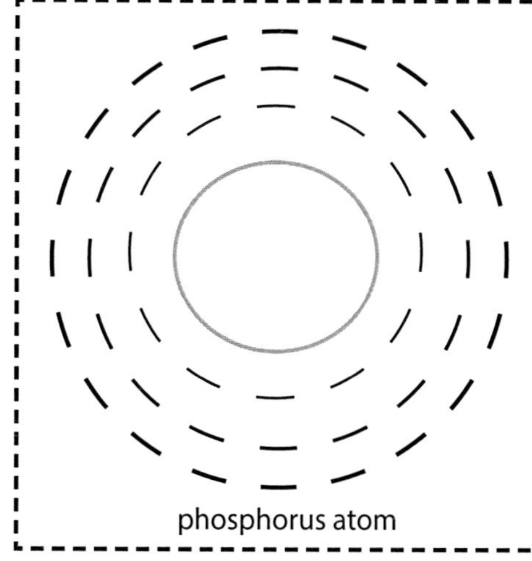

phosphorus atom

Phosphorus Phenomenon

____

____

____

____

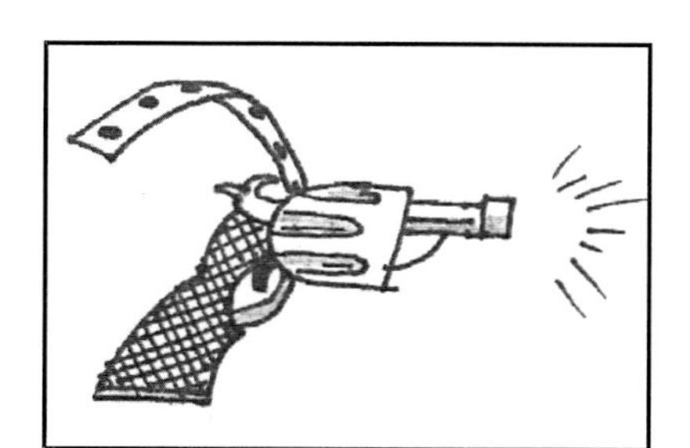

Group 15

P

Phosphorus

NAME ______________________________ DATE ________________

For my notebook

Obnoxious Seagulls - Group 16

Think back to all the molecules you have learned about so far. They all had one element in common. What was it? Did you say oxygen?

The first element in Group 16 is Oxygen.

Symbol: O

Best Known For: There is a lot to tell you, so I better get started.

O is in water and O is in sand.

Plants let O out through their leaves, and we breathe O in.

Without O, fires don't burn.

O protects Earth from the sun's radiation.

Liquid O, when combined with liquid H, makes rocket fuel.

O is essential to life as we know it.

O is the third most common element in the universe.

O is the most common element in Earth's crust and in the ocean.

O is the second most common element in the air we breathe.

We take O for granted on Earth, but if people ever live on the moon, it could be a problem. O is one thing you cannot live without.

The next element in Group 16 is Sulfur (SUL-fer).

Symbol: S

Best Known For: When S combines with oxygen or with hydrogen, it's so Stinky.

S is an important part of gunpowder.

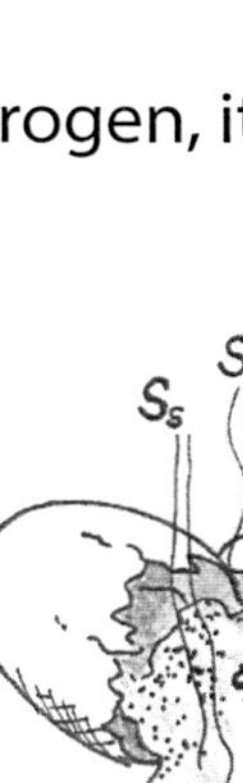

The Greek poet Homer wrote about S nearly 2,800 years ago. He said it was important for keeping pests away.

When they speak about fire and brimstone in the bible, S is brimstone.

Good places to find S are eggs, volcanoes, fireworks, and matches.

Obnoxious Seagulls - Group 16 Lab #1: That's Not My Egg You're Cooking, Is It? - instructions

CAUTION: This lab involves use of a heat source and boiling water. Only parent/instructor should operate the heat source and handle the boiling water.

Materials:

- One egg for each student + one extra egg
- Clear glass
- Heat source and pan to hard-boil the eggs
- Water
- Timer
- Salt, if you plan on eating the eggs
- Knife, for cutting the peeled eggs in half

There is not a lab sheet for this lab.

Aloud: Have you ever eaten a hard-boiled egg with a green ring around the yolk? In a raw egg, the yolk is yellow and the egg white is clear or whitish, so where does a green ring come from when an egg is cooked? There is sulfur in the white of the egg along with iron, another element, in the yolk. The green ring is caused when the iron of the yolk combines with the sulfur of the white. It occurs if the egg is overcooked.

Procedure:

1. Crack a raw egg into a glass. Have everyone look at the egg to check for green around the yolk. There will not be any.
2. Put the rest of the eggs into a pan and cover them with water. Cook the eggs. When the water begins to boil, let them cook for 30 minutes. The goal is to overcook them. When the time is up, drain the water from the pan and let the eggs cool.
3. When they have cooled, peel them and cut them in half. Peel the white off the yolk and look at the outside of the yolk that was nested inside the white. The green that you see is a sulfur-iron compound that formed during the cooking process. The green compound is harmless, and the eggs are safe to eat.

Instructor's Note:

- Sulfur, by itself, has no noticeable smell. The gas that smells like "rotten eggs" is hydrogen sulfide (H_2S). It is what you smell when eggs truly are rotten and is sometimes the result of a rebellious digestive system ☺. The sulfurous smell you may notice after a match is struck is sulfur dioxide (SO_2), the burning combination of sulfur on the match and oxygen in the air.

Obnoxious Seagulls - Group 16 Lab #2:
Bubble Trouble - instructions

Materials:
- Lab sheet, pencil
- Hydrogen peroxide - small bottle
- One raw potato
- Sink or work bucket

Aloud: Have you ever had hydrogen peroxide poured on a cut? It probably hurt, didn't it? Do you remember how it bubbled? Those bubbles are pure oxygen gas coming from the hydrogen peroxide. A hydrogen peroxide molecule has two hydrogen atoms and two oxygen atoms in it. When it bubbles, it is turning into water and oxygen. Cool, isn't it? Let's experiment and see if we can make some bubbles.

Before we do that, though, can you tell me why we use hydrogen peroxide to wash cuts? Is it to kill the germs? Is it the germs that make the hydrogen peroxide bubble?

Procedure:
1. Complete the hypothesis portion of the lab sheet.
2. Pour hydrogen peroxide over your hand on an area of skin with no cuts or scrapes on it. (It should **not** bubble.)

Aloud: If hydrogen peroxide bubbles when it meets germs, it should have bubbled when you poured it on your skin. Whether you know it or not, even if you just washed that part of your skin, you have germs on it.

If it is not the germs that cause hydrogen peroxide to bubble, what does? The chemical that makes the hydrogen peroxide bubble is called catalase (CAT-uh-layss). Catalase is in your blood. When you pour hydrogen peroxide on a cut, the catalase in your blood makes the hydrogen peroxide bubble. Catalase is also in potatoes. When you pour hydrogen peroxide on a piece of potato, will you see bubbles?

Procedure continued:
3. Cut the potato in half. Over the sink, pour hydrogen peroxide on the cut side of the potato. You might have to wait a minute or two before it really starts to bubble.

Instructor's Note:
- You need to make sure the hydrogen peroxide is fresh. Over time, hydrogen peroxide turns into water and oxygen.

Possible Answers:

Hydrogen peroxide does not bubble when poured on your hand unless you have cuts on your hand.

Hydrogen peroxide bubbles when you pour it on a cut potato.

The chemical catalase in the potato makes the hydrogen peroxide bubble.

NAME ______________________________ DATE ________________

Obnoxious Seagulls - Group 16 Lab #2: Bubble Trouble

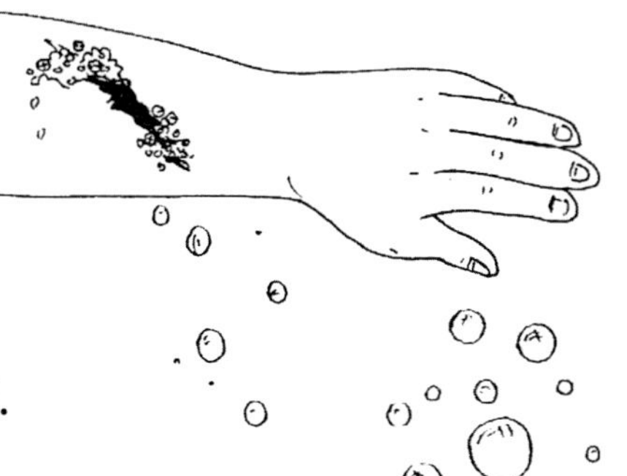

Hypotheses:
What do you think will happen when you pour hydrogen peroxide on your hand?

1) It will bubble.
2) Nothing—I don't have any cuts on my hand.

Do you think hydrogen peroxide will bubble when it comes in contact with germs?

yes no maybe

What do you think will happen when you pour hydrogen peroxide on the potato?

1) Nothing—it's a potato.
2) It will bubble.

Results:
What really happened when you poured hydrogen peroxide on your hand?

What really happened when you poured hydrogen peroxide on the potato?

Discussion and conclusion:
Why did the hydrogen peroxide make bubbles on the potato?

Element Book: Obnoxious Seagulls - Group 16 - instructions

Materials:

- Lab sheet, pencil
- Periodic table found on the inside cover of this book
- "Element Book" (assembled) with student's periodic table
- Scissors
- Glue
- Art supplies - markers, colored pencils, crayons
- Flipbook from Unit 3, for reference

Aloud: What is the atomic number of oxygen? (8) O has the same number of protons as neutrons. How many are there? (8 of each) What is the atomic mass of O? (8 + 8 = 16) What is the atomic number of sulfur? (16) S also has the same number of protons as neutrons. So, what is the atomic mass of S? (16 + 16 = 32) Just like other families, the elements in Group 16 have properties in common because Group 16 elements have the same number of electrons in their outer energy levels.

How many electrons are in the outer energy level for both oxygen and sulfur? Will the answer be the same for both? (yes) All the elements in Group 16 have six electrons in their outer energy levels.

Procedure:

1. On the lab sheet, draw the electrons in the outer energy level around the elemental symbol. (See example below.) Remember, all Group 16 elements have six electrons in their outer energy levels. That is why they are grouped together.
2. Write the atomic number above each symbol.
3. Write the elements' atomic masses. Create an atomic mass table (like the one found on the back of this page) if it will assist your student in calculating the atomic masses. The atomic mass of an element goes right below the symbol and its name in its square. Remember, the atomic mass of an element is the number of protons (atomic number) an atom has + the number of neutrons. Oxygen has 8 neutrons
   Sulfur has 16 neutrons
4. Glue the symbol squares on their appropriate places on the student's periodic table.

Procedure continued: Decorate Group 16 page in "Element Book":

5. Cut out the rest of the items on the lab page.
6. Use information you learned this week about the elements to fill in the trivia boxes for each element, listing characteristics, examples, and interesting facts. Color the items as you wish.
7. Create an atom for each element by drawing in the electrons, protons, and neutrons. Draw each part of the atom a different color. Remember that only two electrons go in the first energy level and up to eight electrons in the second and third energy levels. Both elements should have six electrons in their outer energy levels. (Refer to your flipbook for help drawing the atoms.)
8. On page 8 of your "Element Book," glue the Group 16 label on the tab. Spread out the element labels on the page and glue the other items under their appropriate label.

Answers:

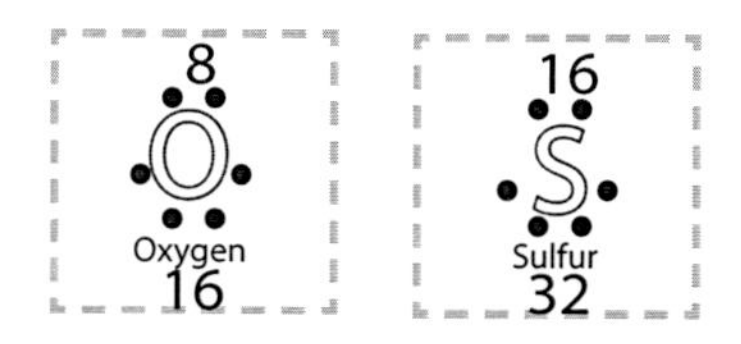

(continued on the back)

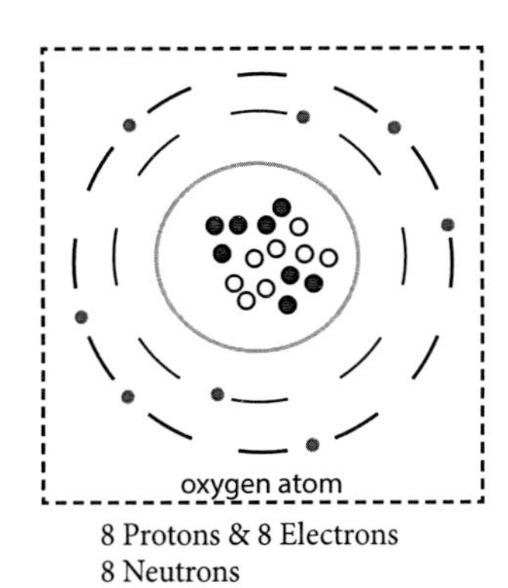

8 Protons & 8 Electrons
8 Neutrons

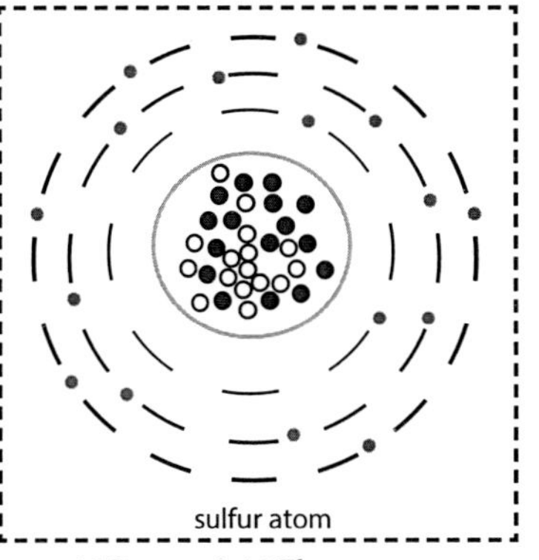

16 Protons & 16 Electrons
16 Neutrons

● = protons
○ = neutrons
• = electrons

Atomic mass table

Element's Symbol	Atomic Number = Number of Protons	+	Number of Neutrons	=	Atomic Mass
O	8	+	8	=	16
S	16	+	16	=	32

Outrageous oxygen wisdom: O is in water and sand, plants give off O, animals breathe it, fires won't burn without it, protects Earth from the sun's radiation, liquid O (when combined with liquid H) makes rocket fuel, is essential to life, third most common element in the universe, most common element in Earth's crust and in the ocean, second most common element in the air.

The stinky on sulfur: when combines with oxygen or with nitrogen, it smells like rotten eggs; an important part of gunpowder; Homer wrote about S keeping pests away; it's referred to as "brimstone" in the bible; found in eggs, volcanoes, fireworks, and matches.

NAME ________________________________ DATE ____________________

Element Book: Obnoxious Seagulls - Group 16

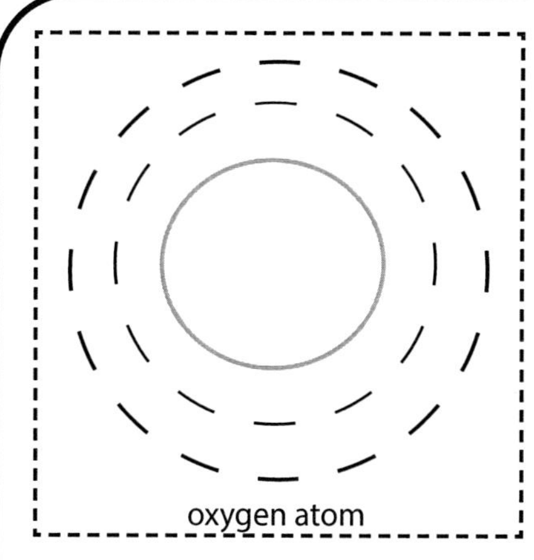
oxygen atom

Oxygen

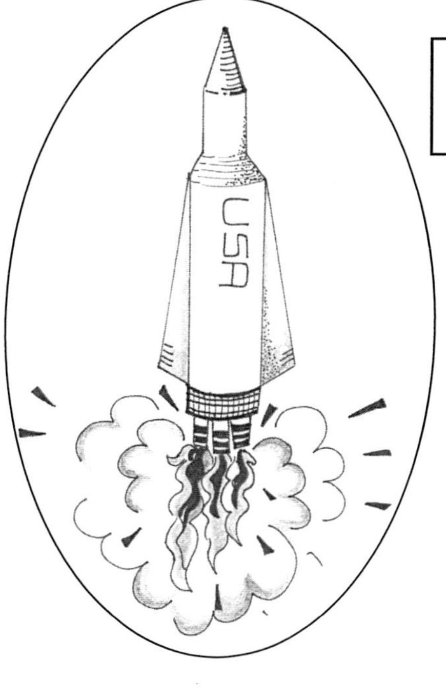

Group 16

Outrageous Oxygen Wisdom

____

____

____

____

____

____

Sulfur

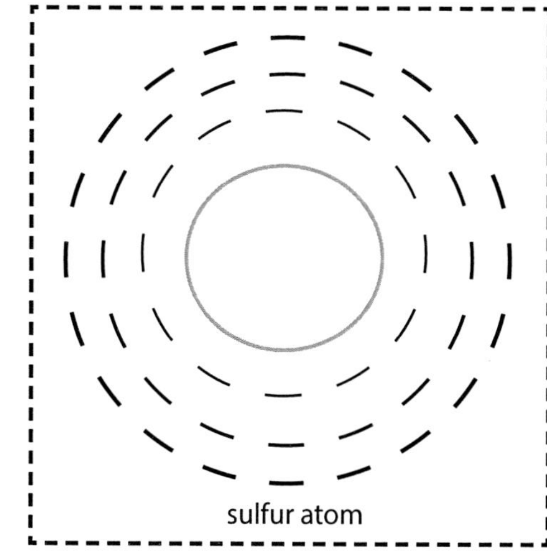
sulfur atom

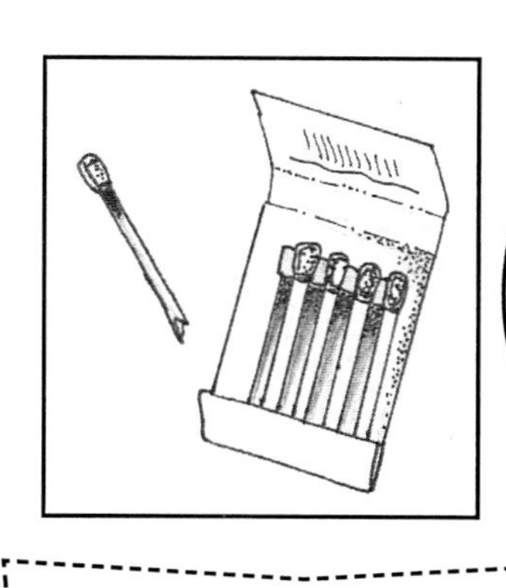

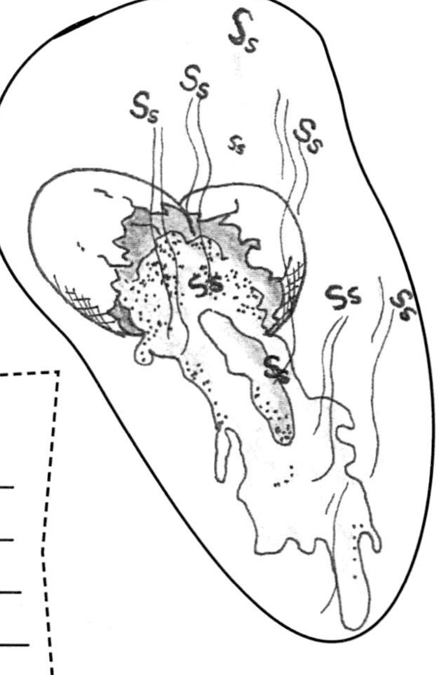

The Stinky on Sulfur

____

____

____

____

____

NAME ______________________________ DATE ________________

For my notebook

Frequently Clever - Group 17

The two elements you will meet today from Group 17 are very reactive. That means that when one of their atoms is around a different kind of atom, they like to link together and form a molecule.

The first element in Group 17 is Fluorine (FLOOR-een).

Symbol: F

Best Known For: If you feel like you have met F before, think "tough teeth." Is it coming back to you? That's right—fluoride treatments, fluoride in your toothpaste, fluoride rinse.

F is also in Teflon pans. That is the black stuff frying pans are coated with so foods won't stick.

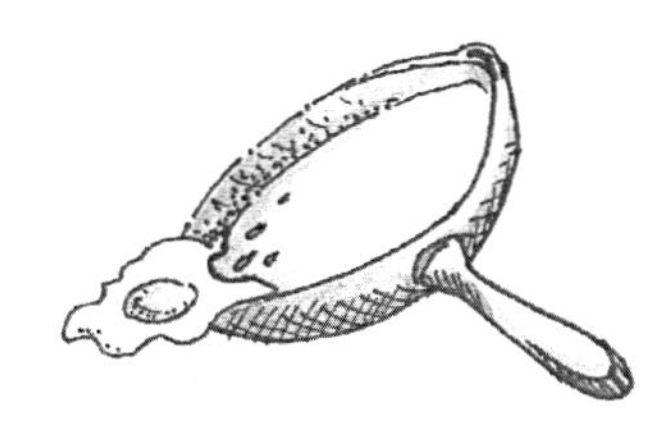

Elemental F is so reactive, almost anything placed in the path of a stream of fluorine gas will spontaneously burst into flames! That includes things like glass and steel.

The second element in Group 17 is Chlorine (CLOR-een).

Symbol: Cl

Best Known For: Clean swimming pools. If you have ever smelled bleach, then you have smelled Cl. Cl is good at whitening clothes and getting rid of yucky stuff in your pool. When Cl is put in pool water, you can smell it. It doesn't smell like something you would want to eat, does it? But sodium (Na) + Cl = salt, and think of how good NaCl makes food taste. **Cl**everly interesting, isn't it?

Cl is used to make paper, like the nice white paper this book is printed on.

Most gases are colorless. Cl is one of the few colored gases. Cl is a pale yellow gas. In bleach, Cl whitens things. But when Cl is a gas floating around all by itself, it's yellow!?!

Frequently Clever - Group 17 Lab #1: Dancing Drops - instructions

CAUTION: This lab uses bleach. Bleach is poisonous and the fumes are toxic. The bleach should only be handled by the parent/instructor.

Materials:

- Lab sheet, pencil
- ½ cup Bleach
- Dark food color
- ½ cup Water
- Glass
- Eyedropper

Aloud: **Bleach has a chemical in it that contains the element chlorine. The name of this chemical is <u>sodium hypochlorite</u> (SO-dee-uhm hye-po-CLOR-ite). Chemicals containing chlorine are used to clean swimming pools, clothes, and sometimes they are used to clean water as a part of making it safe to drink. What do you think will happen if Cl is added to colored water?**

Procedure:

1. Complete the hypothesis portion of the lab sheet.
2. Drip three drops of dark food color into the glass.
3. Pour ½ cup of water into the glass.
4. Draw a labeled picture of the colored water in the glass on the lab sheet.
5. With the eyedropper, carefully drip bleach into the glass until the color disappears.* How many drops of bleach did you need? Record the number on the lab sheet.
6. Draw a labeled picture of what the water in the glass looked like as bleach was added.
7. Drip drops of food color into the water bleach mix. What happens? (The drops of color "dance" and disappear in the bleach/water mix).
8. Draw a labeled picture of what the water in the glass looked like as food color was re-added.

Instructor's Notes:

- Encourage your student to take his time and draw and color detailed pictures of each step in this lab. Drawing and labeling steps of an experiment reinforces good observation skills, organization skills, and the scientific method. It doesn't matter if your student can draw well, but his drawings should be as detailed as possible.
- * The age of the bleach, brand, and location it was bottled can affect its strength. In limited cases, 75 to 100 drops were needed before the color disappeared.

NAME ______________________________ DATE ____________________

Frequently Clever - Group 17 Lab #1: Dancing Drops

Hypothesis:

What do you think will happen to the color of the solution when bleach is added?

1.) The color will disappear.

2.) Nothing will happen.

3.) The color will get darker.

Results:

How many drops of bleach did you use? ___________

Draw a series of labeled pictures below, showing what happened in this experiment. Start with a drawing of the colored water. Next, draw a picture of the water after the bleach was added. Last, draw a picture of what the water looked like when you re-added the food color.

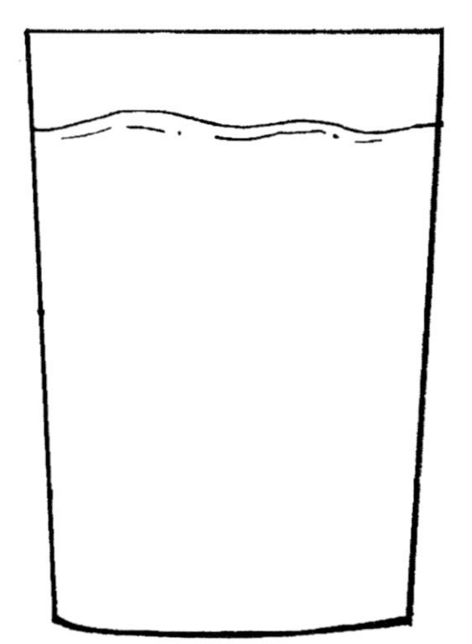

Frequently Clever - Group 17 Lab #2: The Tooth, the Whole Tooth, and Nothing but the Tooth - instructions

This lab takes about six days to complete, but it is a great demonstration of the importance of brushing teeth.

Materials:

- Lab sheet, pencil
- One 4.6 oz tube of fluoridated toothpaste
- One white raw egg
- Vinegar
- A glass
- Water
- Toothbrush
- Spoon
- Colored nail polish
- Plastic wrap

Aloud: How many times has a grown-up told you to brush your teeth? They want you to use toothpaste when you do it, too, don't they? Okay, so, it is pretty gross when someone has stuff in his teeth. But why don't your parents save the money they use to buy toothpaste and buy toys instead? What does toothpaste do and why does it need to have <u>fluoride</u> in it? Let's experiment and find out if fluoridated toothpaste helps make your teeth stronger.

Chicken eggs are made of calcium. Your teeth are, too. You are going to put ½ of a chicken egg in fluoridated toothpaste for 5 days. Then you are going to soak the entire egg in vinegar for 10 to 12 hours. Vinegar will soften an untreated eggshell in 10 to 12 hours and will dissolve it in 24 hours. Will the fluoride bond with the calcium to make a stronger shell (or in your case, stronger teeth)? If it does, the vinegar will not soften the eggshell.

Procedure:

1. Check the egg for cracks. Do not use a cracked or damaged egg.
2. Gently clean the egg with soap and water. Let the egg dry.
3. Paint an X on one end of the egg with colored nail polish. Let the nail polish dry.
4. Complete the hypothesis portion of the lab sheet and record the first observation of the egg.
5. Squeeze the entire tube of toothpaste into the glass.
6. Put the egg, marked side down, in the glass. Do not let the egg touch the bottom of the glass. The toothpaste should cover half of the egg.
7. Cover the glass with plastic wrap.
8. Let the egg sit for 5 full days. Do not refrigerate the egg.
9. Take out the egg and wash the toothpaste off the egg and out of the glass.
10. Record the second observation of the egg.
11. Let the egg dry overnight.
12. The next day, put the egg back in the clean glass.
13. Pour vinegar into the glass so that it covers the egg.
14. Take the egg out 10 to 12 hours later. Be careful, the egg should be quite soft on half of its shell.
15. Record the final observation of the egg and complete the lab sheet.

NAME ______________________________ DATE ____________________

Frequently Clever - Group 17 Lab #2: The Tooth, the Whole Tooth, and Nothing but the Tooth

Hypothesis:

What do you think is going to happen to the side of the egg that sits in fluoridated toothpaste for 5 days? Will the fluoride bond with the calcium to make a stronger shell?

Yes No I don't know

Results and Observations:

Describe the egg before you treated it with fluoride.

____

____

How long did you leave the egg in the toothpaste?____________

Describe the egg after you treated it with fluoride. Does it look different?

____

____

How long did you leave the egg in the vinegar? ____________

Describe the egg after it had soaked in the vinegar.

____

____

Discussion and Conclusion:

What happened to the egg and why is it important for you to brush regularly with fluoridated toothpaste? ______________________________

____

____

Draw a picture showing what your tooth would look like if you did this experiment on it.

Element Book: Frequently Clever - Group 17 - instructions

Materials:
- Lab sheet, pencil
- Periodic table found on the inside cover of this book
- "Element Book" (assembled) with student's periodic table
- Scissors
- Glue
- Art supplies - markers, colored pencils, crayons
- Flipbook from Unit 3, for reference

Aloud: What is the atomic number of fluorine? (9) How many protons does F have? (9) F has 10 neutrons, so what is the atomic mass of F? (9 + 10 = 19) What is the atomic number of chlorine? (17) How many protons does Cl have? (17) Cl has one more neutron than proton. So, what is the atomic mass of Cl? (17 + 18 = 35) Just like other families, the elements in Group 17 have properties in common because Group 17 elements have the same number of electrons in their outer energy levels.

How many electrons are in the outer energy level for fluorine and chlorine? Well, oxygen and sulfur have 6, so the number is 6 + 1 = 7. The elements in Group 17 have 7 electrons in their outer energy levels.

Procedure:
1. On the lab sheet, draw the electrons in the outer energy level around the elemental symbol. (See example below.) Remember, all Group 17 elements have seven electrons in their outer energy levels. That is why they are grouped together.
2. Write the atomic number above each symbol.
3. Write the elements' atomic masses. Create an atomic mass table (like the one found on the back of this page) if it will assist your student in calculating the atomic masses. The atomic mass of an element goes right below the symbol and its name in its square. Remember, the atomic mass of an element is the number of protons (atomic number) an atom has + the number of neutrons. Fluorine has 10 neutrons
Chlorine has 18 neutrons
4. Glue the symbol squares on their appropriate places on the student's periodic table.

Procedure continued: Decorate Group 17 page in "Element Book":

5. Cut out the rest of the items on the lab page.
6. Use information learned about the elements this week to fill in the trivia boxes for each element, listing characteristics, examples, and interesting facts. Color the items as you wish.
7. Create an atom for each element by drawing in the electrons, protons, and neutrons. Draw each part of the atom a different color. Remember that only two electrons go in the first energy level and up to eight electrons in the second and third energy levels. Both elements should have seven electrons in their outer energy levels. (Refer to your flipbook for help drawing the atoms.)
8. On page 9 of your "Element Book," glue the Group 17 label on the tab. Spread out the element labels on the page and glue the other items under their appropriate label.

Answers:

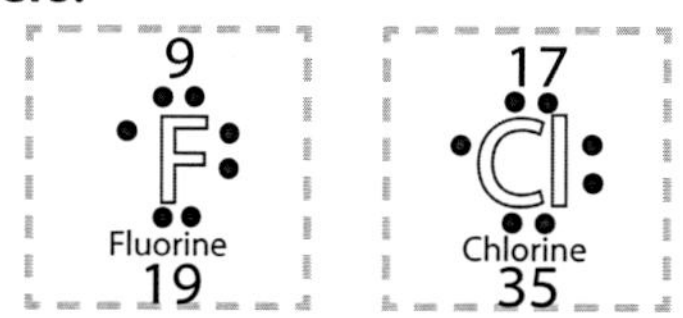

(continued on the back)

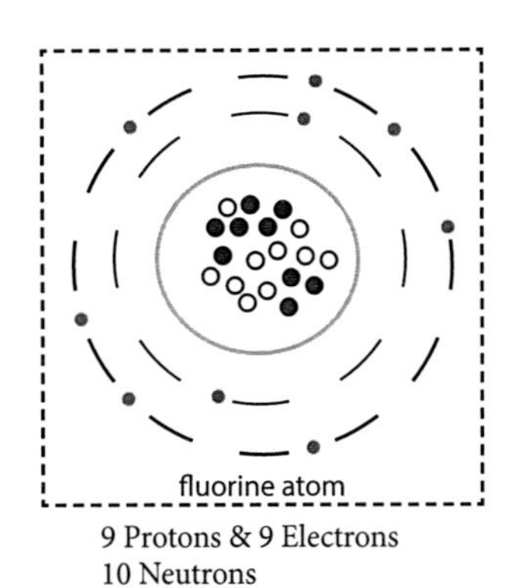

9 Protons & 9 Electrons
10 Neutrons

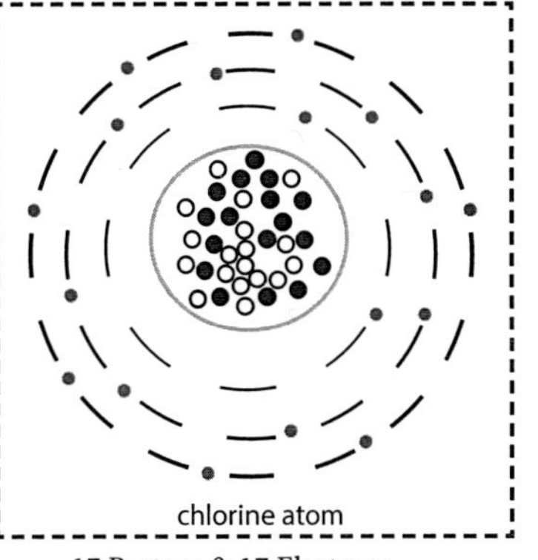

17 Protons & 17 Electrons
18 Neutrons

● = protons
○ = neutrons
• = electrons

Atomic mass table

Element's Symbol	Atomic Number = Number of Protons	+	Number of Neutrons	=	Atomic Mass
F	9	+	10	=	19
Cl	17	+	18	=	35

Fantastic fluorine facts: Found in fluoride treatments, toothpaste, fluoride rinses, and Teflon pans. Very reactive, almost anything placed in the path of a stream of fluorine gas will spontaneously burst into flames.

Clever chlorine data: Used for cleaning swimming pools, found in bleach, noxious (harmful) odor, combines with Na to make salt, used to make paper, pale yellow gas.

NAME ________________________________ DATE ____________________

Element Book: Frequently Clever - Group 17

Fluorine

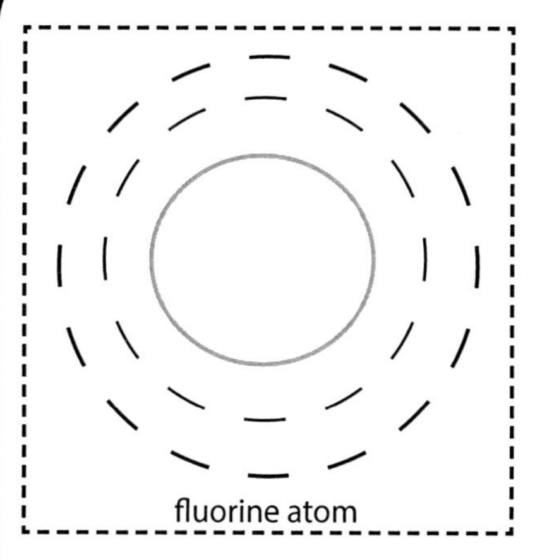
fluorine atom

Fantastic Fluorine Facts

____

____

____

____

Group 17

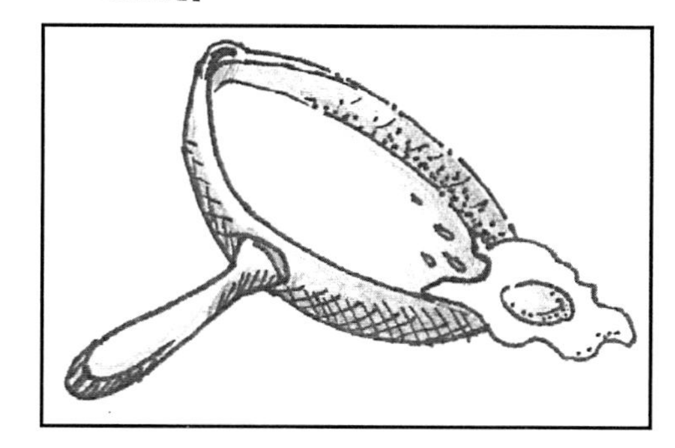

Chlorine

Clever Chlorine Data

____

____

____

____

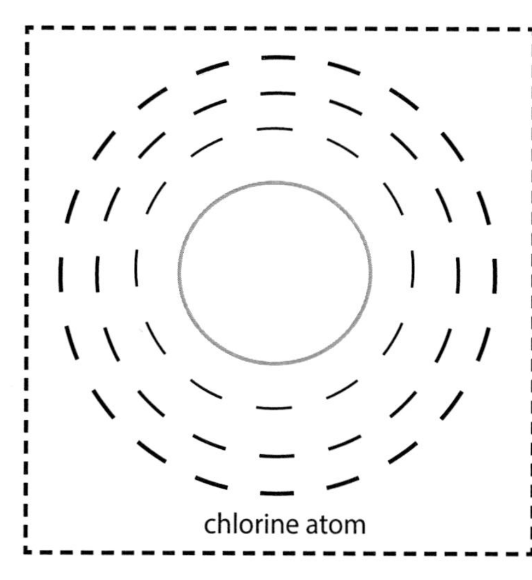
chlorine atom

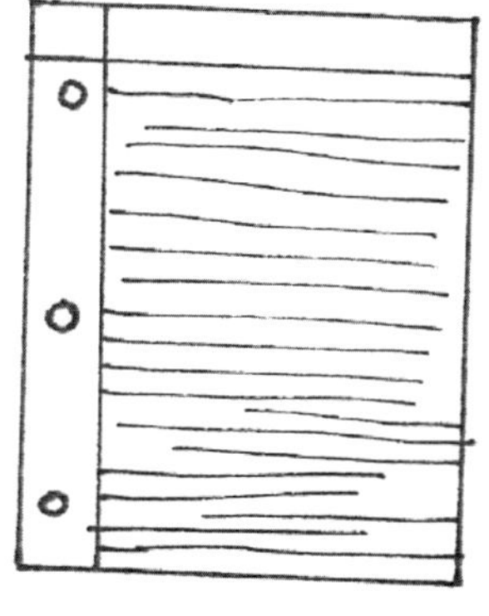

NAME ________________________________ DATE ____________________

For my notebook

He Never Argues - Group 18

It is time to meet the three most common elements from the last column of the periodic table. Many of their friends have already met you, and they feel a little left out.

The first element in Group 18 is Helium (HEE-lee-um).

Symbol: He

Best Known For: He is the gas used to fill up balloons when you want them to float. Have you ever heard someone talk after they have breathed in some He? It makes them sound like Donald Duck. **He**e **He**e **He**e.

When you are looking at stars, you are looking at He.

One in every 10 atoms in the universe is He.

On Earth, He is made deep within the earth. It seeps up from the ground with natural gas. He is so light, though, that most of it escapes into space.

He is the second most common element in the universe.

He is the sixth most common gas in Earth's atmosphere.

The second element in Group 18 is Neon (NEE-on).

Symbol: Ne

Best Known For: Ne is best known for neon lights. Neon lights are the lights in the tubes that glow. Neon gas is in neon tubes, making them glow.

Ne is the fourth most common element in the universe.

Ne is the fifth most common gas in Earth's atmosphere.

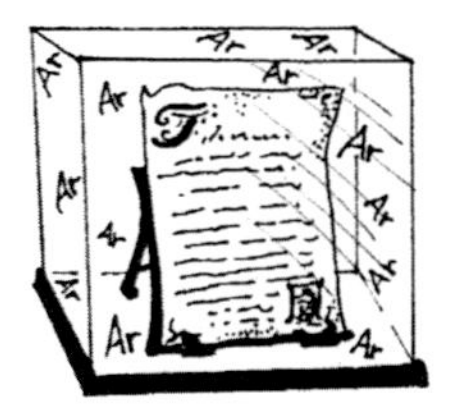

The third element in Group 18 is Argon (AR-gon).

Symbol: Ar

Best Known For: Ar is the third most common gas in Earth's atmosphere. Move your hands through the air. Your hands are hitting Ar. **Ar**e you impressed? Have your ever even heard of Ar before today?

Standard light bulbs (incandescent) were filled with Ar.

Ar is a gas used by museums to preserve old manuscripts and by wine makers to keep wine from turning bad.

He Never Argues - Group 18 Lab: The Incredible Shrinking Balloon - instructions

Materials:

- Lab sheet, pencil
- Two helium-filled* balloons (Many grocery stores sell helium balloons. You only need one for the lab; the extra one is in case it pops.)
- Flexible cloth tape measure (or use a string and a ruler)
- Freezer
- Timer

Aloud: What do you think will happen if you put a helium-filled balloon in the freezer? Will it be the same size after it gets really cold? Do you think the temperature affects the amount of space helium gas takes up? Maybe helium atoms are like people—when they get cold they like to snuggle up with each other.

Procedure:

1. Complete the hypothesis portion of the lab sheet.
2. Measure the width around a helium-filled balloon with the string or tape measure.
3. Record the length.
4. Put the balloon in the freezer.
5. Set the timer for 30 minutes.
6. Do you think the balloon will be the same size when it comes out of the freezer? Record your prediction.
7. After 30 minutes, take the balloon out of the freezer.
8. Measure the balloon with the tape measure.
9. Record the length.
10. Let the balloon warm back up for 30 minutes.
11. Measure the size of the balloon.
12. Record the length.

Possible Answers:

Discussion/Conclusion

Yes, cold helium gas takes up less space than room-temperature helium. The balloon shrinks in size. When helium is warmed up, the molecules spread out and take up more space. The balloon grows in size.

Instructor's Note:

- * Some stores add a compound called "hi-float" to helium balloons (to keep them floating longer) which could negatively affect the results. We recommend helium balloons without hi-float for this lab.

NAME ______________________________ DATE ________________

He Never Argues - Group 18 Lab: The Incredible Shrinking Balloon

Hypotheses:

Check the one you think is correct.

- ☐ I think the balloon will be a different size for all three measurements —before it gets cold, when it is cold, and after it warms back up.
- ☐ I think the balloon will be the same size for all three measurements.
- ☐ I think the balloon will be the same size when it is warm, but different when it is cold.

If you think the balloon will be a different size when it is cold, do you think it will be

SMALLER or BIGGER.

Results:

Measurements:

Before being put in the freezer, the balloon was ____________ around.

Right after the balloon came out of the freezer, the balloon was __________ around.

After the balloon warmed back up for 30 minutes, the balloon was _________ around.

Discussion and Conclusion:

Does cold helium gas take up less space than room-temperature helium gas?

YES or NO

What happens when the helium warms back up?

Element Book: He Never Argues - Group 18 - instructions

Materials:

- Lab sheet, pencil
- Periodic table found on the inside cover of this book
- "Element Book" (assembled) with student's periodic table
- Scissors
- Glue
- Art supplies - markers, colored pencils, crayons
- Flipbook from Unit 3, for reference

Aloud: What is the atomic number of helium? (2) How many protons does He have? (2) He has the same number of protons as neutrons, so what is the atomic mass of He? (2+2=4) What is the atomic number of neon? (10) How many protons does Ne have? (10) Just like He, Ne has the same number of protons as neutrons, so what is the atomic mass of Ne? (10+10=20) What is the atomic number of argon? (18) Ar has 22 neutrons, so what is the atomic mass of Ar? (18+22=40)

Now I am going to tell you something strange about Group 18—not all the elements have the same number of electrons in their outer energy levels! That's inconvenient, you might say. And just when we were getting used to the rules! Well, in science there are sometimes exceptions to the rules. But the outer energy levels of helium, neon, and argon do have SOMETHING in common even if it is not the NUMBER of electrons. Can you guess what that might be? They all have outer energy levels that are ALL FILLED UP.

Procedure:

1. For group 18, it will be easier to have students start with making their atoms so they can see the number of electrons in the outer energy level for each atom. Create an atom for each element by drawing in the electrons, protons, and neutrons. Draw each part of the atom a different color. Remember that only two electrons go in the first energy level and up to eight electrons in both the second and third energy levels. Helium has two electrons in its outer (and only) energy level, neon has eight electrons in its outer (second) energy level, and argon also has eight electrons in its outer (third) energy level. Help students to realize that the outer energy levels are filled for all three elements (one more electron and another energy level would be needed).
2. On the lab sheet, draw the electrons that are in the outer energy level around the elemental symbol. (See example on the back.)
3. Write the atomic number above each symbol.
4. Write the elements' atomic masses. Create an atomic mass table (like the one found on the back of this page) if it will assist your student in calculating the atomic masses. The atomic mass of an element goes right below the symbol and its name in its square. Remember, the atomic mass of an element is the number of protons (atomic number) an atom has + the number of neutrons.
   Helium has 2 neutrons
   Neon has 10 neutrons
   Argon has 22 neutrons
5. Glue the symbol squares on their appropriate places on the student's periodic table.

Procedure continued: Decorate Group 18 page in Element Book:

6. Cut out the rest of the items on the lab page.
7. Use information you learned about the elements this week to fill in the trivia boxes for each element, listing characteristics, examples, and interesting facts. Color the items as you wish.
8. On page 10 of your "Element Book," glue the Group 18 label on the tab. Spread out the element labels on the page and glue the other items under their appropriate label.

(continued on the back)

Answers:

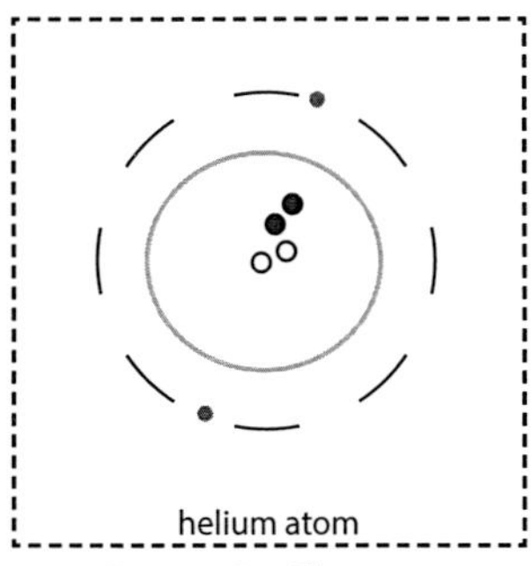

2 Protons & 2 Electrons
2 Neutrons

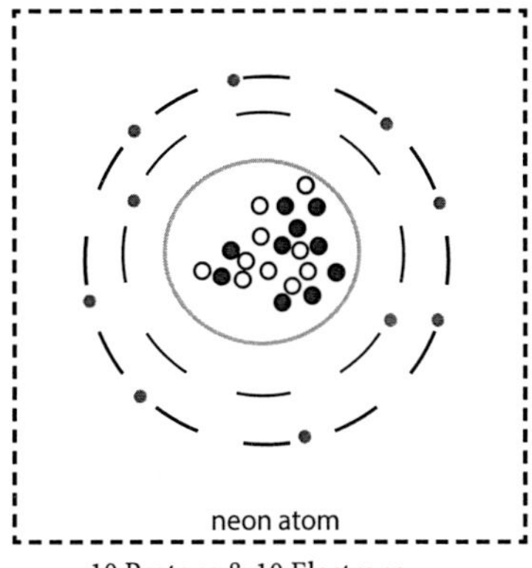

10 Protons & 10 Electrons
10 Neutrons

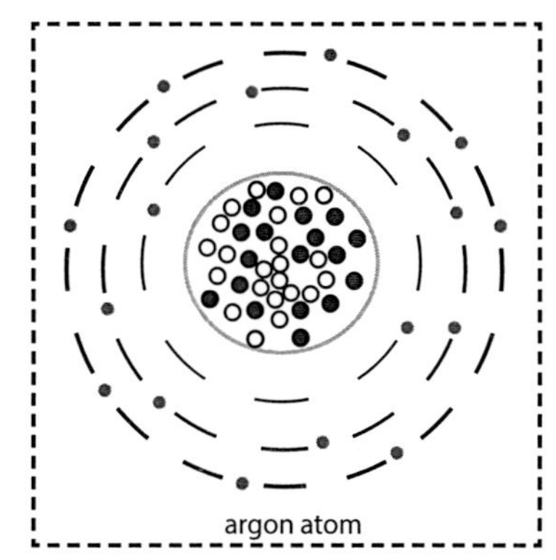

18 Protons & 18 Electrons
22 Neutrons

● = protons
○ = neutrons
• = electrons

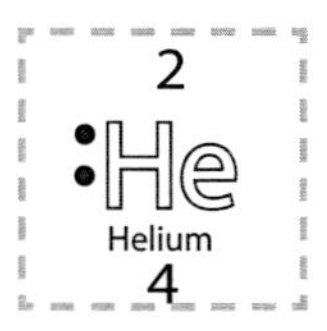

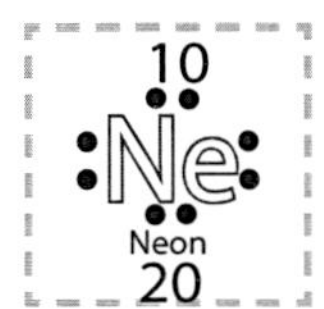

Atomic mass table

Element's Symbol	Atomic Number = Number of Protons	+	Number of Neutrons	=	Atomic Mass
He	2	+	2	=	4
Ne	10	+	10	=	20
Ar	18	+	22	=	40

Hilarious helium hype: Gas used in helium balloons, found in stars, found in one in every 10 atoms in the universe, found deep in the ground on Earth, seeps up and escapes, very light, second most common element in the universe, sixth most common gas in Earth's atmosphere.

Nifty neon: Neon lights and tubes, fourth most common element in the universe, fifth most common gas in Earth's atmosphere.

Argon particulars: Third most common gas in Earth's atmosphere, found in light bulbs, used by museums to preserve old manuscripts, used by wine-makers to keep wine from turning bad.

NAME ______________________ DATE ______________

Element Book: **He Ne**ver **Ar**gues - Group 18

Helium

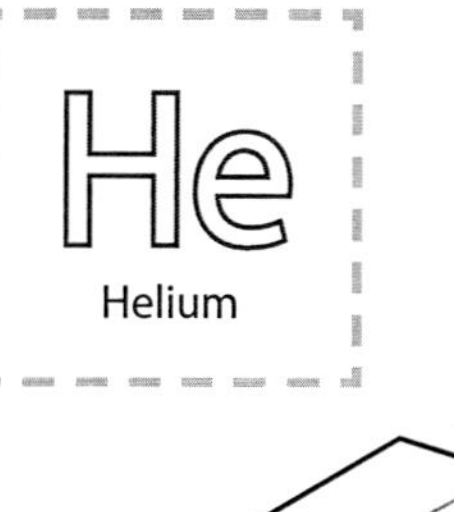

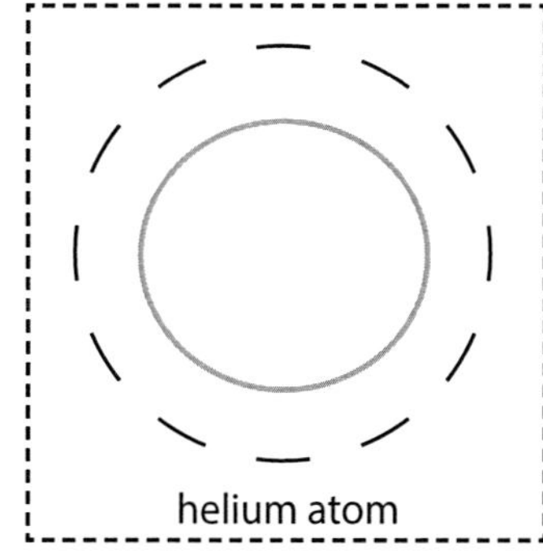

Hilarious Helium Hype

____

____

____

____

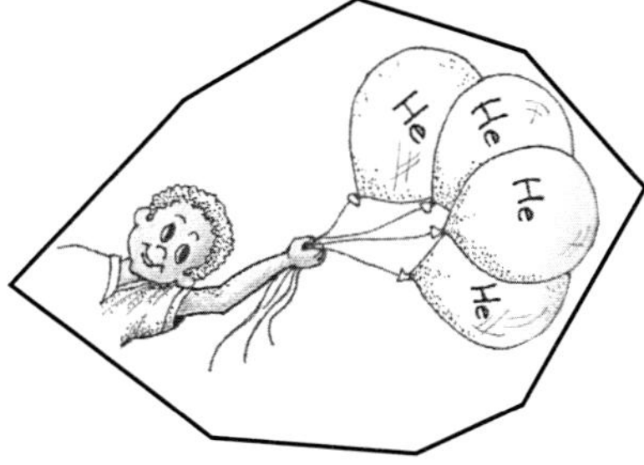

Neon

Nifty Neon

____

____

____

____

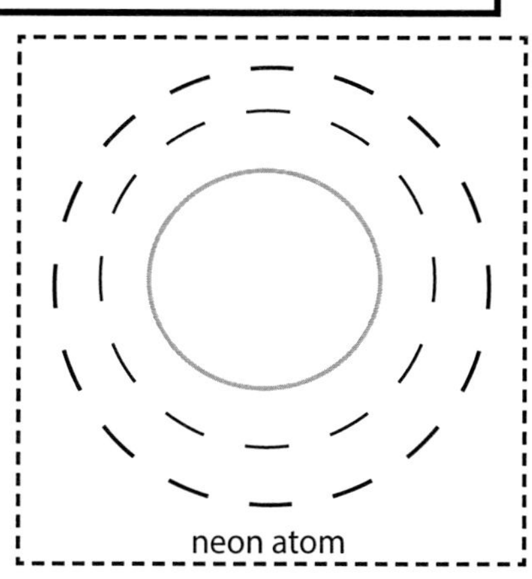

Argon

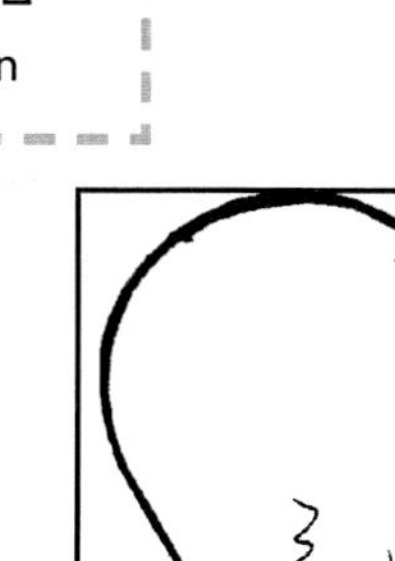

Group 18

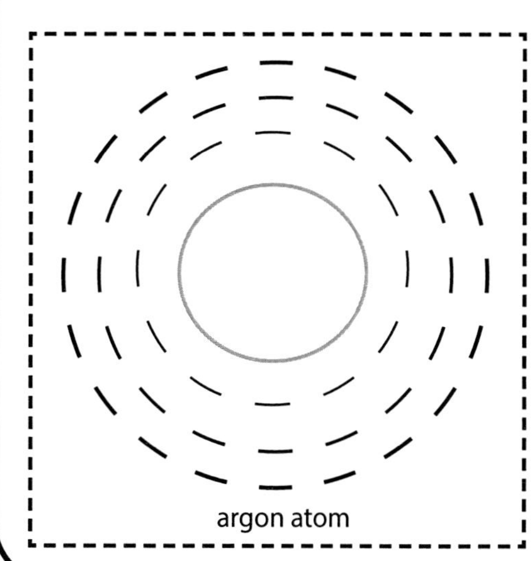

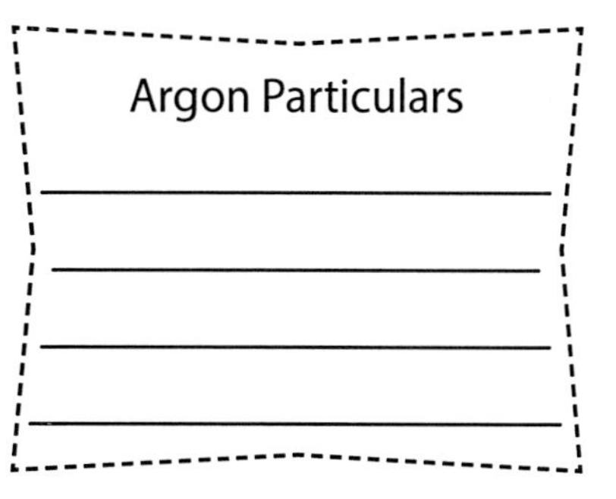

The Chemist's Alphabet Applied - Crossword Vocabulary Review

Write the element name for each symbol. Bonus: One element you added to Your Periodic Table is not on the puzzle. Which one is it?

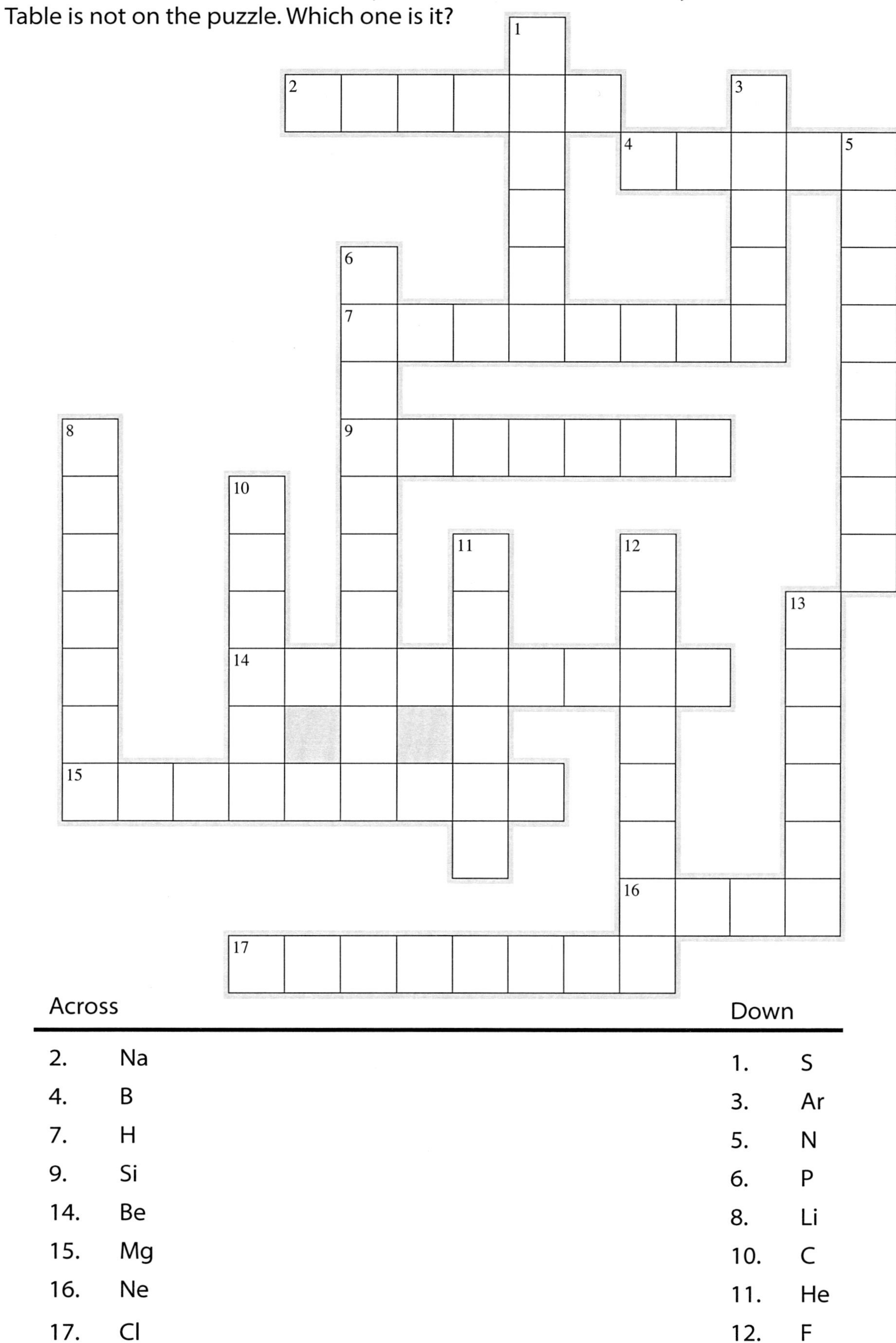

Across		Down	
2.	Na	1.	S
4.	B	3.	Ar
7.	H	5.	N
9.	Si	6.	P
14.	Be	8.	Li
15.	Mg	10.	C
16.	Ne	11.	He
17.	Cl	12.	F
		13.	O

Unit 5

Molecules Rule

Making Molecules

First you have
Protons
Electrons
Neutrons
Yeah!

They make atoms
Come what may.
Then those atoms
You know what
They do?

They play together
With electrons
And make
A molecule.

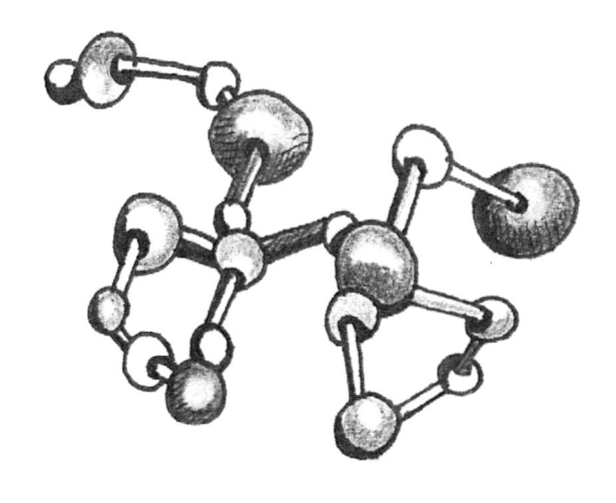

NAME ______________________ DATE ______________

For my notebook

Putting It All Together

Do you like to play with friends? Atoms love to play with their friends. As you know, atoms are made from protons, electrons, neutrons, Yeah!! When atoms play together, they NEVER share their protons or neutrons. Instead, atoms share, swap, and play with each other's electrons.

When atoms get together and play with each other's electrons, they link together. The places where they link together are called bonds. A molecule is a group of two or more atoms bonded together. When atoms bond with each other to make molecules, they use the electrons from their outer energy levels. You can look at Your Periodic Table to see how many electrons each atom uses when they play with their friends. The electron dots you made around each element symbol tells you how many electrons an element plays with when they make molecules. These electrons like to pair up when they play. That's why when atoms play with other atoms, they link together on the sides where they have single electrons. Two single electrons make a pair. Atoms love to make pairs. So you can think of electrons as the glue that holds things together. You could even say electrons are the glue that holds all the things in the universe together.

Putting It All Together Activity: Make a Molecule Puzzle - instructions

Materials:

- Make a Molecule puzzle sheet
- Lab sheet, pencil
- One piece of 8 ½ x 11 card stock
- Glue
- Scissors

Aloud: Molecules are two or more atoms bonded together. Today you will be putting together four Make a Molecule puzzles. You will combine atoms of sodium, chlorine, hydrogen, and oxygen to make the molecules. The electrons in the outer energy levels of the atoms are on each puzzle piece because these are the electrons that like to play the Make a Molecule game. When you make the molecules, don't forget that electrons like to pair up when they play.

Procedure:

1. Glue the worksheet onto card stock. (Alternatively, you could use a copy machine to copy the puzzle page onto a piece of card stock.)
2. When this has dried, cut out the puzzle pieces.
3. Use the puzzle pieces to make the molecules indicated on the lab sheet. Glue or tape the completed puzzles to the lab sheet.

For More Lab Fun: Students can play around with the pieces, bonding different atoms together to make new molecules before gluing or taping them on the page.

Instructor's Notes:

- Help students to notice that where the atoms bond, the electrons make a pair.
- The puzzle pieces for this activity and the elements drawn on "My Periodic Table" have a special name. This type of drawn representation is called an Electron Dot Structure. Electron Dot Structures can be drawn for a single element or for a molecule. This topic will be covered in a later section of this book.

Answers:

Bonus question: Yes, the Na atom could take the place of one of the H atoms in the three molecules with hydrogen.

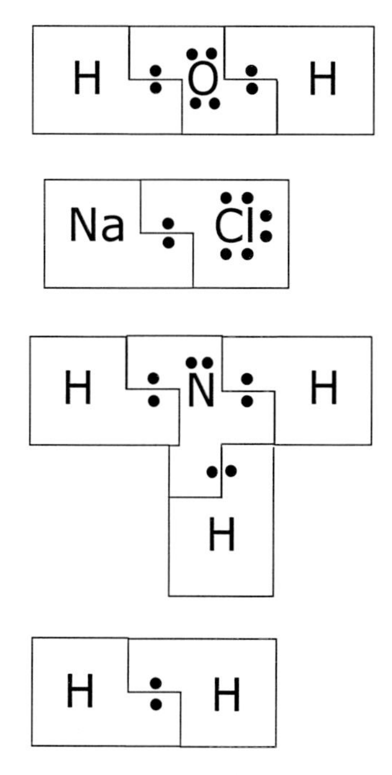

NAME ______________________________ DATE ________________

Putting It All Together Activity: Make a Molecule Puzzle - Puzzle sheet

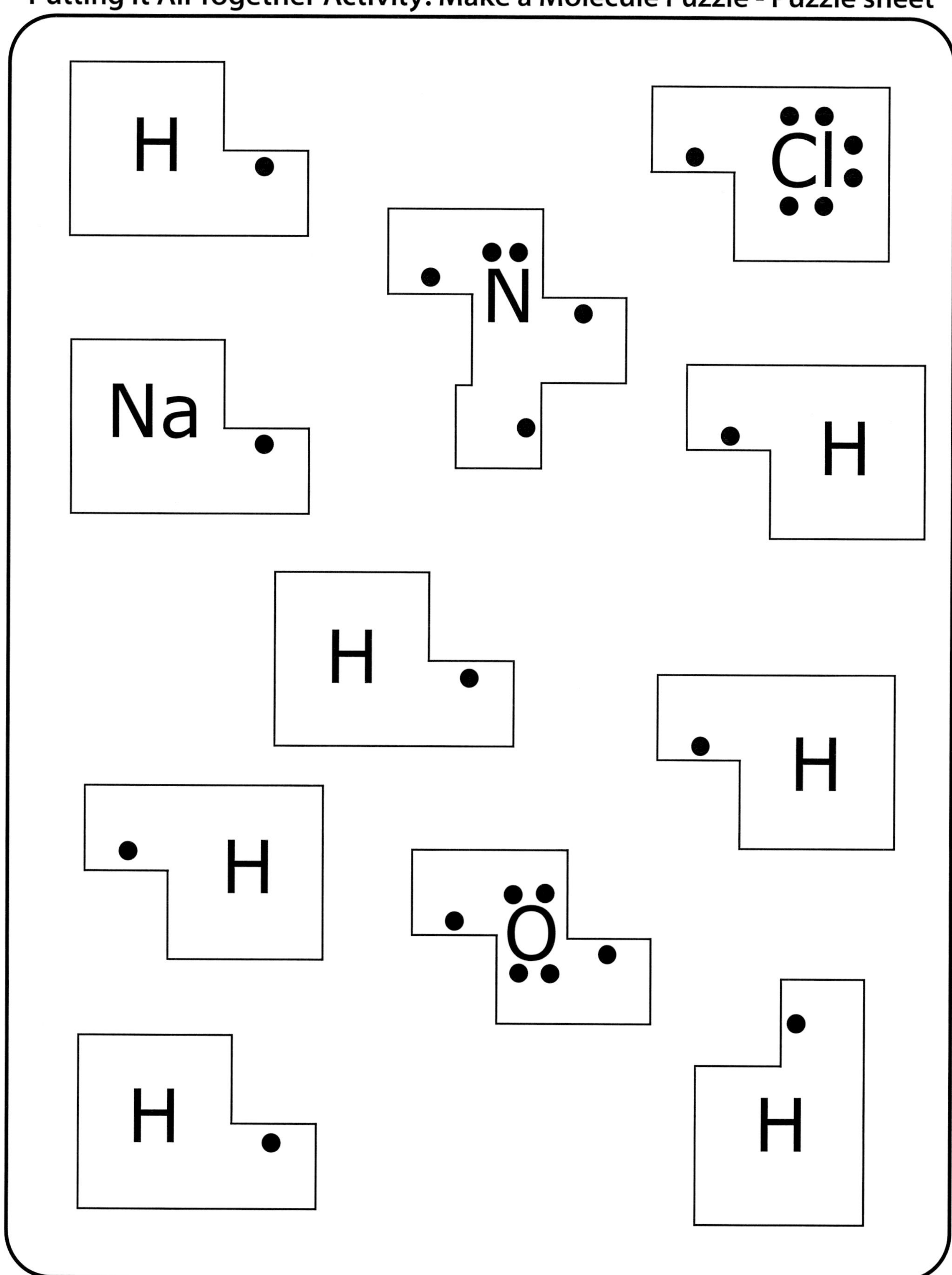

NAME ______________________________ DATE ________________

Putting It All Together Activity: Make a Molecule Puzzle

1. Make the molecule HOH, water.

2. Make the molecule NaCl, salt.

3. Make the molecule ammonia, which has one nitrogen, N, atom and three hydrogen, H, atoms.

4. Make the molecule HH.

Bonus: Sodium, Na, wants to play with nitrogen, N, and oxygen, O, too. Do you think the Na atom could take the place of one of the hydrogen atoms in the three molecules with hydrogen?

Putting It All Together Lab #1: Friendly Gumdrops - instructions

Materials:
- Lab sheet
- One bag of gumdrops (assorted colors)
- Toothpicks
- Colored pencils
- Large work surface

Aloud: First, you learned about atoms. Then you learned that a group of atoms with the same atomic number and the same number of protons in the nucleus, is called an element. Next, you learned that atoms can link together, called bonding, to form molecules. Often molecules are formed from different types of atoms, but sometimes the same type of atoms get together and form a molecule. Now I want you to learn that molecules also like to group together. "The more the merrier" is the motto most molecules have. Most of the time different types of molecules get together. But sometimes all the molecules are the same kind. A group of molecules that are all the same is called a compound. If the group has different kinds of molecules, the group is called a mixture. Now you can show all you have learned using friendly gumdrops that like to get together as much as molecules do.

Procedure:
1. At the top of a large work surface, place one gumdrop to represent one atom.
2. Under the "atom" and to the left, place a handful of the **same color** gumdrops to represent an element.
3. Take two to four of the **same color** gumdrops and use a toothpick(s) to join them. Place these to the right of the "element" to represent a molecule.
4. Take two to four gumdrops with at least one of a **different color** (they can all be different colors) and use a toothpick(s) to join them. Place your second "molecule" next to your first "molecule." Leave a little space between them; don't group them together. Here you are demonstrating that both of these represent molecules because they have atoms that are bonded together even though one has all the same type of atoms and one has different atoms.
5. Now make several (two to four) more "molecules" that are all the same (the same color and number of gumdrops bonded together with toothpicks). The molecules may consist of different atoms or the same type of atom; just make sure that they are all the same molecule. Group these molecules and place them below the first two molecules you made to represent a compound.
6. Now make several more "molecules" (two to four), but this time make them all different (different colors and different numbers of atoms). Group these to the right of the "compound" to represent a mixture.
7. Complete the lab sheet by drawing pictures of your atom-gumdrop creations.

Instructor's Notes:
- Refer to the chart on the back of this page for possible answers.
- Allow students to practice making their own gumdrop elements, molecules, mixtures, and compounds.
- The difference between an element and a molecule is that the atoms in an element are just grouped, whereas in a molecule, they are bonded together (as represented by the toothpicks).
- After making the individual molecules in steps 5 and 6, do not bind the molecules to each other with toothpicks when making a mixture and a compound. Just group the molecules. If you bind the molecules together, you will have a really big molecule and not a mixture or a compound.

(continued on the back)

Possible Answers:

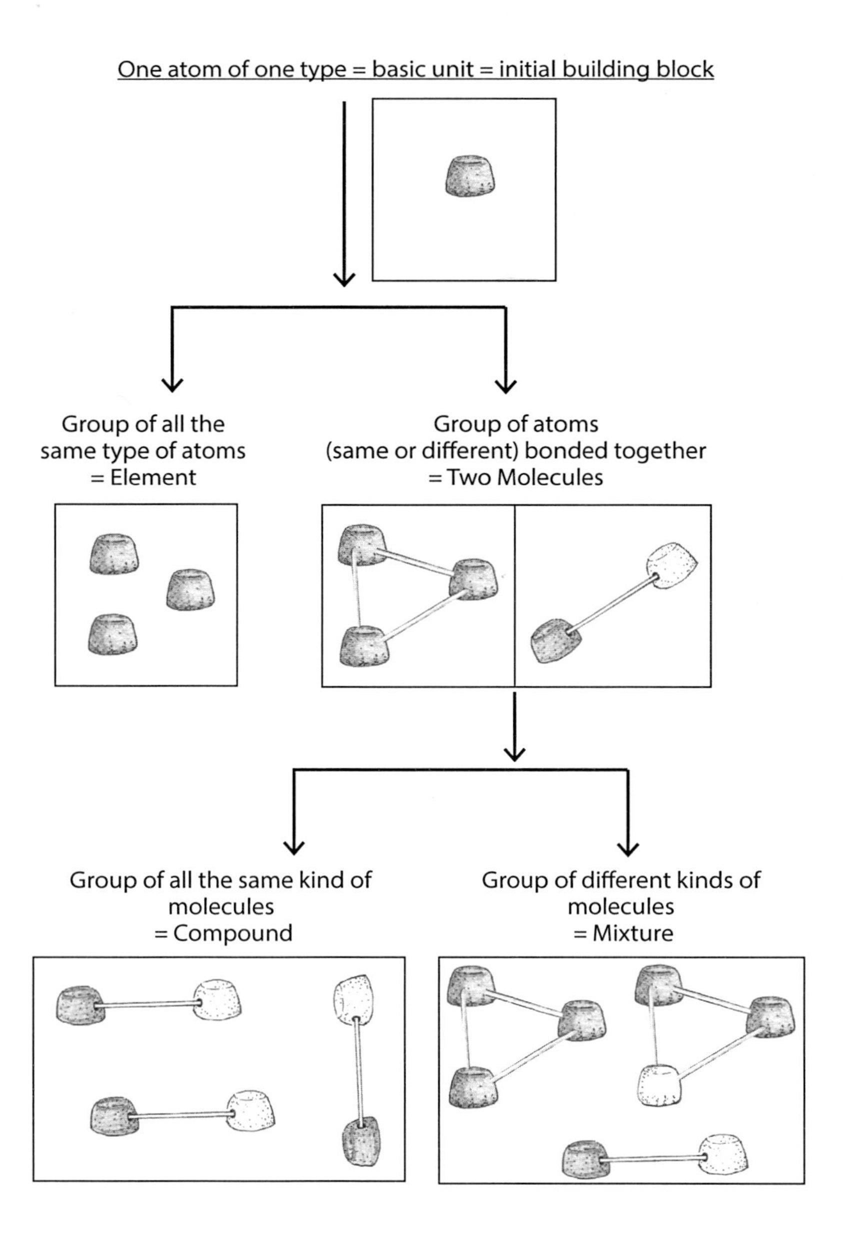

NAME ______________________ DATE ______________

Putting It All Together Lab #1: Friendly Gumdrops

One atom of one type = basic unit = initial building block

Group of all the same type of atoms = element

Group of atoms (same or different) **bonded together** = two Molecules

Group of all the same kind of molecules = compound

Group of different kinds of molecules = mixture

Putting It All Together Lab #2: Mixture or Compound? - instructions

Materials:

- Lab sheet, pencil
- 4 tablespoons of Sugar
- 1 tablespoon of Salt
- 1 tablespoon of Flour
- ½ teaspoon of Pepper
- 1 tablespoon of Peanut butter
- ¼ cup Oil
- Clear glass of water
- Lettuce
- Sliced tomato
- Chopped or sliced carrot
- Salad bowl

Procedure:

1. Today's lab starts with a demonstration for you to do while you are reading the Aloud part. Fill a clear glass half full of water.

Aloud: When atoms bond with each other, they form molecules. A group of all the same type of molecules is called a compound. Water is an example of a compound. A glass of water would be a glass filled with all the same type of molecules—water molecules or H_2O. Therefore, this glass is filled with a compound. What if oil is mixed into this glass of water?

2. Stir ¼ cup of oil into the water.

Aloud: As you can see, the oil and water do not form a compound. When groups of different kinds of molecules are stirred together, it is called a mixture. Now the glass is filled with a mixture of oil and water molecules. Today you are going to look at different things and decide: Is it a compound or is it a mixture?

3. Make a salad. Put lettuce, sliced or chopped carrots, and sliced tomatoes in it.
4. Put each of the following in its own separate small bowl/dish: 1 tablespoon sugar, ½ teaspoon pepper, 1 tablespoon flour, 1 tablespoon salt.
5. Fill in the first part of lab sheet, through question number 5. Help students see that salad is a mixture of clearly defined different things. Sugar, pepper, flour, and salt are each compounds.
6. Set out the peanut butter.
7. Stir 1 tablespoon of sugar into the pepper, 1 tablespoon sugar into the flour, and 1 tablespoon sugar into the salt.
8. Complete the rest of the lab sheet (6 - 9). All of these are mixtures. Sugar + salt and the peanut butter are mixtures that look like compounds because 1.) sugar and salt look alike and 2.) the different types of molecules in peanut butter are very well mixed.

Instructor's Note:

- Pepper can look like a mixture because sometimes you can see different colors (especially if you use course cracked pepper). But it is a compound made of different-colored peppercorns (pepper molecules).

NAME ______________________ DATE ______________

Putting It All Together Lab #2: Mixture or Compound?

(Circle the correct answer.)

1. Salad is a mixture compound.
2. Sugar is a mixture compound.
3. Pepper is a mixture compound.
4. Flour is a mixture compound.
5. Salt is a mixture compound.

6. Peanut butter is a mixture compound.
7. Sugar + pepper is a mixture compound.
8. Sugar + salt is a mixture compound.
9. Sugar + flour is a mixture compound.

NAME ______________________________ DATE ________________

For my notebook

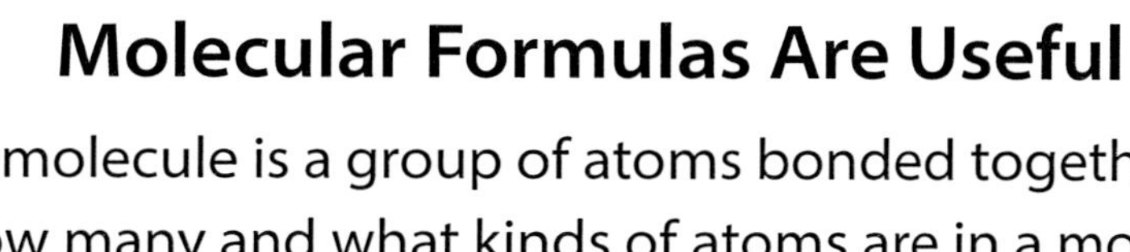

Molecular Formulas Are Useful

A molecule is a group of atoms bonded together. How can you tell how many and what kinds of atoms are in a molecule? For that, you need a <u>molecular formula</u>. That may sound like a complicated term, but molecular formulas are really quite simple to understand.

You have seen water written H_2O. H_2O is one way of writing the molecular formula for water. Two other ways to write the formula are HOH and HHO. This means that there are two hydrogen atoms and one oxygen atom in every molecule of water. As you can see, molecular formulas do not tell you *how* the atoms are bonded together within the molecule. Molecular formulas just tell you the *amount* and *type* of atoms present. H_2O, HOH, and HHO are all correct ways to write the molecular formula for water. H_2O is the most common way to write the molecular formula for a good reason, though. What would the molecular formula look like for a larger molecule, such as sugar? One molecule of sugar has 45 atoms in it. There are 12 carbon atoms, 22 hydrogen atoms, and 11 oxygen atoms. Using the atomic symbols for these atoms, the formula could be written like this:

CCCCCCCCCCCCHHHHHHHHHHHHHHHHHHHHHHOOOOOOOOOOO

Did you count the atoms to make sure I put the correct number in? It would be easy to make a mistake. Think about how many different ways these atoms could be arranged when writing this molecule. When you look at a molecule that is as big as sugar, you know why chemists write it $C_{12}H_{22}O_{11}$.

The process for writing molecular formulas is simple. Write the symbol for each element. After you write the symbol, count how many atoms of that element are present. Write that number next to the symbol in <u>subscript</u>, like the 2 in H_2O. Then go on to the next element. If there is only one atom of an element in a molecule, you do not need to write the 1 after the symbol. That is why there is no 1 written after the O in H_2O.

By looking at the molecular formula, you can tell how many atoms of an element are in a molecule. H_2SO_4 is the molecular formula for <u>sulfuric acid</u>. It has two hydrogen atoms, one sulfur atom, and four oxygen atoms. Molecular formulas are useful and quite easy, don't you think?

NAME ______________________________ DATE ________________

Molecular Formulas Are Useful: Worksheet

1. **C_2H_6O** is a molecular formula.

    How many carbon atoms does it have? ________

    How many hydrogen atoms does it have? ______

    How many oxygen atoms does it have? ________

2. **True or False** There are two atoms of magnesium in **Mg_3N_2**.

3. **True or False** There are two sodium atoms in **Mg_3N_2**.

4. **True or False** The molecule **NNHHHHOO**, has four hydrogen atoms in it.

5. Write the molecular formula for **NNHHHHOO** using subscript.

    ____

6. Write the molecular formula for caffeine, the molecule below, using subscript.

    CCCCCCCCHHHHHHHHHHNNNNOO

    ____

Challenge: The molecule that makes plants green is called <u>chlorophyll</u>. It is really big. Can you write its molecular formula using subscript?

CCCCCCCCCCCCCCCCCCCCCCCCCCCCCCCCCCCHHHHHHHH
HHHHHHHHHHHHHHHHHHHHOOOOONNNNMg

____

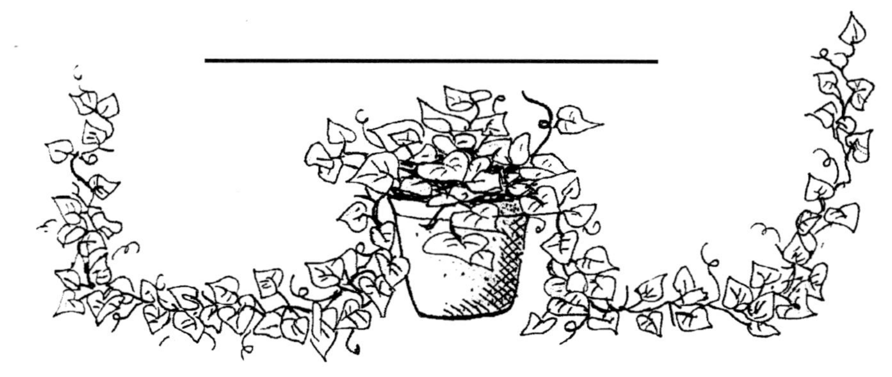

Molecular Formulas Are Useful: Worksheet - answers

1. 2,6,1
2. False—there are three
3. False—there are no sodium (Na) atoms
4. True
5. $N_2H_4O_2$
6. $C_8H_{10}N_4O_2$
7. $C_{35}H_{28}O_5N_4Mg$

   (There are several types of chlorophyll molecules. This is chlorophyll c2.)

Molecular Formulas Are Useful Lab: The Celery Blues - instructions

This lab takes about a day to observe the final results.

Materials:
- Lab sheet, pencil
- A stalk of celery (with leaves)
- A glass
- Blue food coloring
- Water

Aloud: Do you ever help water the plants? Did you know that the water you put at the bottom goes all the way up to the leaves at the top of the plant? How do plants do that? That's like you drinking from your feet! When plants do this, it is called capillary action. Capillary action happens because water molecules are sticky. They stick to each other and they stick to other types of molecules.

Plants have long, thin, straw-like tubes going from their roots to their tops. These straw-like parts are called xylem. A plant's xylem is made from a special type of molecule called cellulose. The water molecules like sticking to cellulose molecules. Cellulose must be fun to play with. Water molecules travel up plants using capillary action. The water molecules travel up and up and up, sticking to the cellulose and to each other, and dragging more and more water molecules along with them.

Let's prove that plants do this. Put some celery in blue water and see what happens. Will the water travel up the celery stalk? If the blue water goes all the way up the stalk to the leaves, we will know that plants can drink water from their "feet."

Procedure:
1. Draw a picture of the celery on the lab sheet.
2. Put 2 inches of water into the glass.
3. Put enough food coloring in the water to make it a very dark blue color.
4. Cut the bottom end off the celery.
5. Put the celery into the water.
6. After the celery has been in the water for 1 hour, draw a picture and label it.
7. After the celery has been in the water overnight or at the end of the day, draw a picture and label it.
8. When you are done letting the celery sit in the water, take it out and cut the stem halfway up.
9. Draw a picture of the cross-section of the celery and label it.

Possible Answers:

Discussion/Conclusion:

The blue color at the top of the celery and in the middle of the stalk (above where the celery was sitting in the water) shows that water put at the bottom of the celery goes to the top.

The xylem are clearly visible in the celery's cross-section. Those are the straw-like circles you can see.

Bonus: The capillaries in your body are tiny blood vessels. The capillary action that happens in them helps your heart move the blood through your body. Your blood is mostly water.

NAME ______________________________ DATE ________________

Molecular Formulas Are Useful Lab: The Celery Blues

Observations:

Drawings and Labels of the Celery Stalk

Before	After 1 hour	At the end
Drawing of the xylem in the cross-section of the celery stalk:		

What happened to the blue water? Did the celery drink it up with its "feet"?

Bonus: You have capillaries in your body. What do you think happens in them?

NAME ______________________________ DATE ____________________

For my notebook

Drawing Lessons

Molecules are really small. If you tried to draw a life-sized molecule, it would be impossible, because like atoms, a molecule is too small for a person to draw. One molecule is so small it would be impossible for the mite that is on a flea that is on a hair that is on a cat to draw. Even a really big molecule is only a little larger than an atom.

Chemists do have a method for drawing molecules, though. This method is called the Electron Dot Method. The molecule puzzles you made use this method. These kinds of drawings are called Electron Dot Structures. You already know how to draw Electron Dot Structures. Look at Your Periodic Table—the atomic symbols with the electrons you drew around them are Electron Dot Structures for those elements.

When you put those atoms together, they make molecules. To draw a simple molecule using the Electron Dot Method, there are some easy rules to follow:

1.) Draw one atom and its electrons as it appears on Your Periodic Table.

2.) Do not start with hydrogen (he is so polite that he always lets everyone else go first).

3.) Draw the rest of the atoms, one at a time, showing their electrons sharing with the atom you drew first.

4.) Atoms share electrons on the sides where they have a single electron.

Let's make water and see how this works.

Drawing Lessons: Worksheet - instructions

Materials:

- Lab sheet, pencil
- "My Periodic Table"
- Blank paper for student to practice drawings
- Chalkboard, dry erase board, or paper for drawing demonstration

This is a scripted lesson. Read the instructions below aloud to students as you draw the sample on a chalkboard.

Aloud: Instructions for drawing the Electron Dot Structure of water, H_2O:

1. **Look at Your Periodic Table and find oxygen.**
2. **On your paper, write down the symbol, O, including the electrons from its outer energy level, as drawn on Your Periodic Table. This is the Electron Dot Structure for the element oxygen.**

Your drawing should look like this:

3. **Next, look at Your Periodic Table for hydrogen.**
4. **Write down the symbol for one of the hydrogen atoms with its electron, as drawn on Your Periodic Table. Draw this next to oxygen with its electron. Remember, when atoms play together, they like to pair up and link together on the side where they have a single electron. When hydrogen and oxygen atoms form bonds, they share one electron from each atom. It doesn't matter which single electron of oxygen you choose.**

5. **There is another hydrogen to draw. Do the same thing on the opposite side for the second hydrogen atom.**
6. **There will be only two electrons on each of the four sides of oxygen. When drawing Electron Dot Structures for molecules, there are always only two electrons on each of the four sides of a molecule.**
7. **When I drew the molecule, I did not take any electrons away from either atom. That is why when I drew the molecule, oxygen still has its six electrons around it and each hydrogen atom still has its one electron.**

The finished water molecule looks like this:

Lab sheet

8. Complete the lab sheet, having students create their own Electron Dot Structures.

(continued on the back)

Answers:

1. Methane, CH_4

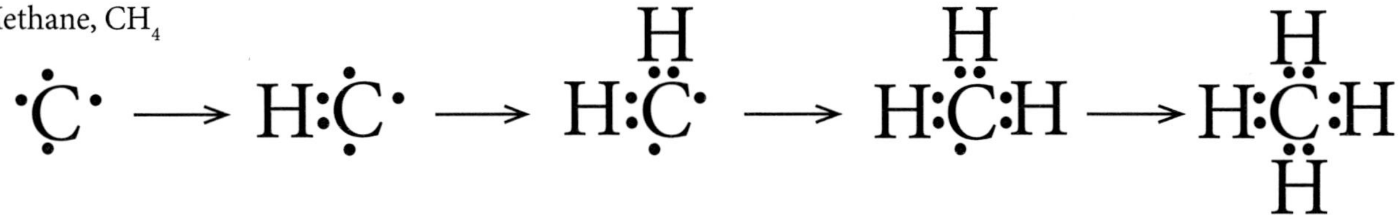

$$C \longrightarrow H:C \longrightarrow H:C(H) \longrightarrow H:C:H\,(H) \longrightarrow H:C:H\,(H)(H)$$

2. Ammonia, NH_3

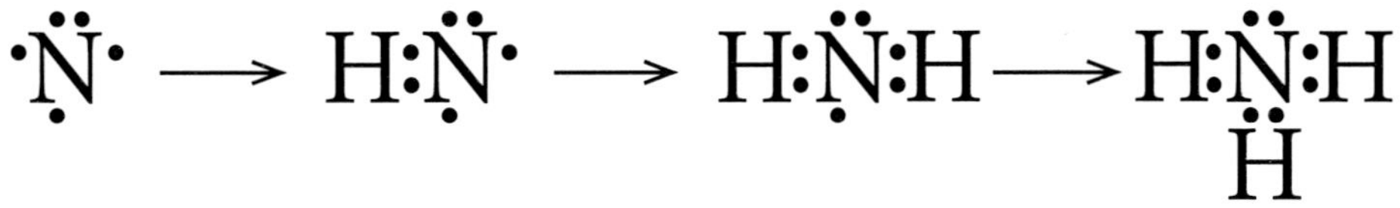

$$N \longrightarrow H:N \longrightarrow H:N:H \longrightarrow H:N:H\,(H)$$

2. Fluoride, HF

$$F \longrightarrow H:F$$

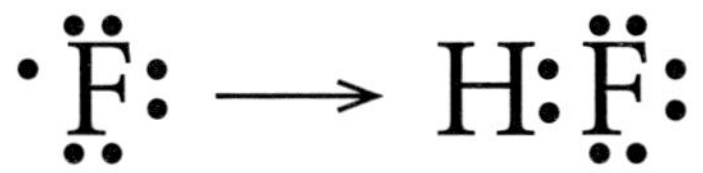

Instructor's Notes:

- The answers above show all the steps of the drawings for your reference. Students should not be expected to show all these steps if they can arrive at the correct answer more directly.
- The placement of the electrons around each symbol can vary. Just be sure the electrons bond in pairs.

NAME ______________________ DATE ______________

Drawing Lessons: Worksheet

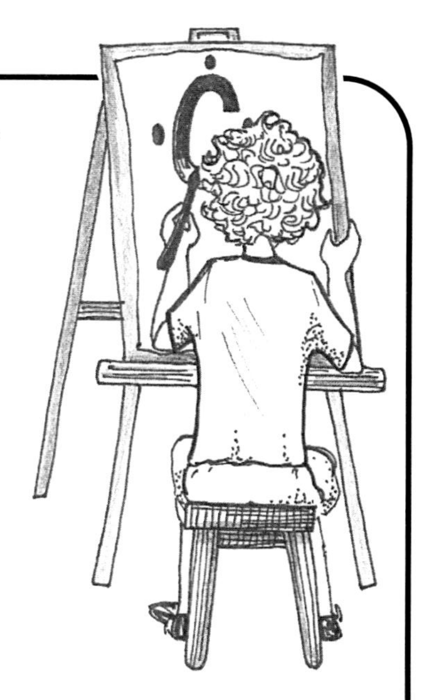

1. Now it is your turn. Draw the Electron Dot Structure for the molecule methane, **CH_4**. Start with carbon.

2. Draw the Electron Dot Structure for an ammonia molecule, **NH_3**. Start with nitrogen.

3. Draw the Electron Dot Structure for hydrogen fluoride, **HF** . Start with fluorine.

Drawing Lessons Lab: Capillary Action In Action – instructions

Materials:

- Lab sheets (two pages), pencil
- One square of WHITE paper towel with NO designs on it
- Water (about a cup)
- Three shallow dishes (the saucers that go with teacups work well)
- Tablespoon
- Food coloring
- 1 tablespoon Vegetable oil or olive oil
- Q-tip
- Scissors
- Ruler
- Tape
- A surface to work on that does not stain. If you do not have this, put plastic wrap, aluminum foil, a tray, or wax paper under the saucers and the paper towel strips.

Aloud: When a paper towel touches water, it absorbs the water. If you dip just the edge of a paper towel in water, the water will climb up the paper towel. This is another example of capillary action. Trees are made of cellulose, and paper towels are made from trees. Guess what type of molecule paper towels have in them? Water molecules love to play with the cellulose molecules in paper towels just like they did in celery.

Have you ever seen anyone put fertilizer on a plant? Fertilizer is like a multi-vitamin for plants. The fertilizer dissolves in the water and the water carries the fertilizer up the plant. Fertilizer must find water sticky. When you put the food coloring on a paper towel, the water will carry it up the paper towel just as a plant does with fertilizer. Water must like to play with those molecules, too.

Water molecules are like the kids of the molecular world. They just want to play, play, play, play, play, play all the time. Guess what, though; they don't like to play with everybody. Water loves to play with some molecules, like cellulose and those in food coloring. But there are types of molecules that water does not like to play with at all. Water does its best not to play with molecules it doesn't like.

Do you remember when you stirred oil into water? The oil and water did not really mix together, and after a short while the oil sat on top of the water. What if there is a line of oil on the paper towel? Will that affect how the water moves? The answer to that depends how well water likes to play with oil and how well it sticks to the oil. You will look at two strips of paper towel—one with food coloring on it, and another with food coloring and a line of oil on it. Will the oil on the paper towel make a difference as to how fast the water moves up the paper towel?

Procedure:

1. Cut two strips of paper towel into 2-inch (5 cm) wide strips. Turn the strips vertically and measure 2 inches (5 cm) from the bottom of each strip and make a light pencil mark.
2. On both strips, put a drop (JUST a drop) of food coloring at the spot you measured. Let these strips dry for 10 minutes. Complete the hypothesis portion of the lab sheet while you are waiting.
3. Pour a little bit of oil into a dish. Take a Q-tip and dip it into the oil. On one of the strips of paper towel about 1 inch (2.5 cm) above the food coloring, take the Q-tip with the oil on it and run it all the way across the short length of the paper. You might have to dip the Q-tip in the oil and run it across the paper towel more than once. Make sure the oil saturates the paper towel but does not start to run down into the food coloring or up the paper towel.
4. Measure 2 tablespoons of water into each of the two shallow dishes that do not have oil on them.

(continued on the back)

5. Put the ends of the two strips of paper towel into the water at the end that is nearest to the food coloring drop. Make sure the food coloring drops are not in the water but that the ends of the paper are.
6. As the water is absorbed up the paper towel strips, the water level will drop. Make sure the end stays in the water.
7. When the water has finished traveling to the other end of the paper towel, record your observations.
8. Let the strips dry, then tape them to the back of the lab sheet.

Possible Answers:

The water will climb up the paper towel strips and take the food coloring with it. The oil does affect the movement briefly. There is enough water in the dish that it overwhelms the oil strip, like a dam being breached.

Discussion/Conclusion:

You do observe capillary action in this experiment. Water does not find oil as "sticky" as it does food coloring, water itself, or the paper towel.

Instructor's Note:

- Capillary action results from two forces: cohesion and adhesion. It occurs when the adhesive molecular forces between a liquid and a substance are stronger than the cohesive intermolecular forces inside the liquid.
- The two capillary action labs are included in this unit because they are good labs for demonstrating molecular interaction. They do not segue naturally from molecular formulas or from drawing electron dot structures. These two experiments, do however, fit perfectly in the unit that introduces molecules.

NAME ______________________________ DATE ________________

Drawing Lessons Lab: Capillary Action In Action - page 1

Hypotheses: (circle your answer)

Will the water and food coloring move up the paper towel?

Yes   No   I don't know

Will the oil have any effect?

Yes   No   I don't know

I think the oil will ____________________ the water and coloring.

slow down   speed up   do nothing to

Results / Observations:

What happened when you put the paper towel strips with food coloring into the water? Did the oil have any effect on the movement of the water?

Discussion and Conclusion:

Did you observe capillary action? Do you think water likes to play with oil?

Drawing Lessons Lab: CAPILLARY ACTION IN ACTION - page 2

Paper Towel Strip #1 - no oil

Paper Towel Strip #2 - with oil line

Molecules Rule - Crossword Vocabulary Review

Across

1. Atoms make molecules by linking these.
4. A group of atoms bonded together.
7. An abbreviated way to write the atoms in a molecule. (Two words)
9. One way to draw molecules. (Three words)

Down

2. This happens when water travels up a plant. (Two words)
3. The straw-like parts of plants.
5. A group of different kinds of molecules.
6. How atoms link together in a molecule.
8. A group of all the same kind of molecule.

Unit 6

What's the Matter?

NAME ______________________ DATE ______________

What's the Matter? Poem

Mountains
Aardvarks
Twinkling Stars
Trees
Everything
Really

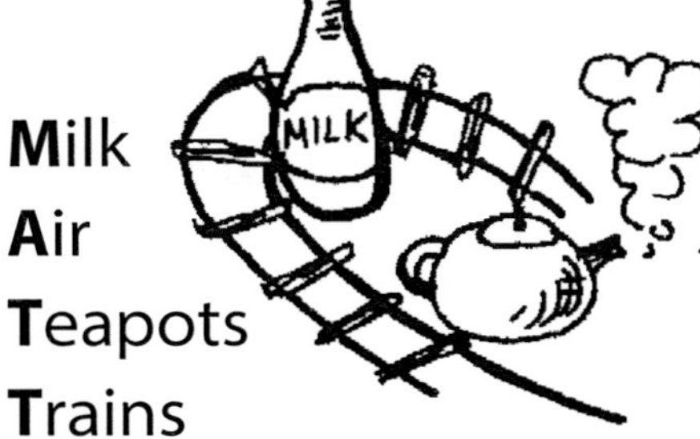

Milk
Air
Teapots
Trains
Everything
Really

Mooing Cows
Angelfish
Tables
Toys
Everything
Really

The ears of a **M**ouse
Art in a house
Toenails of a louse
Spu**T**tering flames to douse
Everything
Really

M
A
T
T
E
R

Matter: An Introduction - Activities - instructions

The introduction to matter begins with three activities—a worksheet, a puzzle, and a poem.

Materials:

- Build a Drop of Water Matter worksheet (page 271)
- Atoms Are the Matter worksheet (page 273)
- What's the Matter? Poem (page 268)
- "My Periodic Table" (assembled in the Element Book)
- Pencil
- Colored pencils
- Blank paper
- Assorted Lego pieces (about 10 to 15 pieces, including one large blue piece and two identical small white pieces)

<u>Activity #1 - Build a Drop of Water Matter</u>

Aloud: A drop of water is a drop of matter. <u>Matter</u> is anything that has mass and takes up space. Look around you; everything you can see and touch is matter. Even the air around you is matter, though you cannot see it or feel it. First, you learned how protons, electrons, neutrons (Yeah!) form atoms. Then you learned about the different types of atoms. Next, you learned that these different types of atoms can link together and form molecules. Now, you are going to learn that when molecules group together, they make matter. On the worksheet, you are going to start at the beginning with protons, electrons, neutrons (Yeah!) and build an important life-giving type of matter: a drop of water. You might need to refer to Your Periodic Table to find some of the answers. As you complete the worksheet, remember that all matter consists of protons, electrons, neutrons, atoms, and molecules.

Procedure: Complete Build a Drop of Water Matter worksheet using "My Periodic Table" for reference.
Part 1 - Write down the total number of protons, electrons, and neutrons in the spaces labeled Part 1.
Part 2 - Draw Electron Dot Structures for each atom of H, hydrogen. O, oxygen has been done for you.
Part 3 - Draw the Electron Dot Structure for each molecule of water. The first one has been done for you.
Part 4 - Draw a drop of water, a drop of matter.

Answers:

Part 1 - Oxygen = 8 protons, 8 electrons, 8 neutrons. Hydrogen = 1 proton, 1 electron, 0 neutrons
Part 2 - H˙ + ·Ọ̈· + ˙H
Part 3 - H:Ọ̈:H + H:Ọ̈:H + H:Ọ̈:H = a drop of water
Part 4 - Draw a drop of water with H_2O molecules

H_2O H_2O H_2O

<u>Activity #2 - Atoms Are the Matter</u>

Aloud: Do you remember who Democritus was? He was the Greek guy who first said that all things were made from atoms. It was not until over 2,000 years after Democritus lived that people started to wonder about atoms again. <u>John Dalton</u> was born in England in 1766. His family was very poor and could only afford to send him to school until he was 11 years old. He was so smart, though, that he began to teach school when he was 12 years old! Can you imagine being a schoolteacher at 12 years old? Later, he became a famous scientist. He believed Democritus was right. In 1803, when he was 37 years old, John Dalton wrote the <u>Atomic Theory of Matter</u>. That is a fancy way to say he figured out some important things about how atoms combine to make matter. Today you are going to pretend that Legos are atoms and parts of atoms that you can put together to make matter. The atoms make bigger parts that are molecules. When you put the larger molecule or Lego clumps together, they make matter.

(continued on the back)

Procedure:

1. First let students play around with the Legos, making "matter." Lay out the Lego pieces you plan on using. Look at the pieces. Think of each piece as a proton, neutron, or electron. Build the "atoms." Put these together to make "molecules." Put the "molecules" together. When the Lego structure has been built, look at the "matter" you have made.
2. On the Atom's Are the Matter worksheet, you will find each part of John Dalton's Atomic Theory of Matter (the highlighted parts). Read the following scripted lesson to students (each statement of the theory is underlined in the script) and have them follow the instructions using Legos and filling in their worksheets. In this demonstration, each single Lego represents one atom.

Aloud: <u>All things are made of atoms</u>. Pretend each Lego piece is an atom. Pick out two Legos. Draw pictures of the two Legos on the worksheet under number 1. You have just drawn two Lego atoms.

<u>Atoms that are the same size and shape as each other are the same type of atom</u>. Choose two Legos that are the same size, shape, and color. These Legos are one type of atom. Draw these two Lego atoms in the column labeled "same" on your worksheet in box 2a.

<u>Atoms that are different from each other are different types of atoms</u>. The Legos that are different shapes, sizes, and colors are not the same type of atoms. Choose two Legos that are different from each other. Draw them in the column labeled "different" in box 2b.

<u>Atoms combine with each other to make molecules</u>. Put some Legos together. (You are going to draw the molecule you make, so you might want to keep your molecule small.) You have made a Lego molecule. Draw your molecule under number 3a.

You have heard a lot about the molecule H_2O. H_2O is water. It has two hydrogen atoms and one oxygen atom. Find one large blue Lego. It will be Lego oxygen. Now, find two white Legos that are smaller than the blue Lego. We will call these white pieces Lego hydrogen. The Lego hydrogen atoms need to be the same as each other. Attach the two white Legos to the blue Lego, not to each other. You have just made Lego water matter. Draw your Lego water under number 3b.

<u>Sometimes when two different types of molecules get together, they switch atoms around and make different molecules. When this happens, it is called a *chemical reaction*</u>.

Now make two different types of molecules from Legos. Draw these on your worksheet in box 4a. Take some Legos off each molecule and switch the pieces around. Draw these new molecules you made in box 4b, demonstrating a Lego chemical reaction.

<u>When chemical reactions happen, atoms cannot be created or destroyed</u>. They switch around to make new molecules. All the same atoms are still there, though. If my dog chewed one of my Lego atoms to pieces, would that be a chemical reaction? No, because atoms are not destroyed or created in chemical reactions. That would just mean I needed to pick up my Legos.

Instructor's Note:

- The use of Legos in this experiment should help those students who are kinesthetic/tactile learners to understand these principles. Chemical reactions are studied in more depth in unit 7.

<u>Activity #3 - What's the Matter? Poem</u>

Aloud: Look at the What's the Matter? poem. This type of poem is called an <u>acrostic</u>. In an acrostic, a letter in each word in the poem spells out another message. The message in our poem is Matter! Add a stanza to this poem by creating an acrostic verse of your own.

Procedure:

1. Complete the last stanza of the poem, creating your own acrostic for MATTER. Any words are correct since everything is matter (except pure energy such as light and heat). The last two lines should read "Everything Really," but creative minds may differ.
2. Use colored pencils to illustrate your stanza.

NAME ______________________________ DATE ____________________

Build a Drop of Water Matter

Part 1: Write the total number of protons, electrons, and neutrons (Yeah!) for each element.

protons +	1	____	____
electrons +	1	8	____
neutrons (Yeah!)	0	____	____
	↓	↓	↓
	atom hydrogen	+ atom oxygen	+ atom hydrogen

Part 2: Draw the Electron Dot Structure for each atom.

H $\quad$ $\cdot\ddot{\underset{\cdot\cdot}{O}}\cdot$ $\quad$ H

Part 3: Make molecules of water.

molecule $\quad$ $H:\ddot{\underset{\cdot\cdot}{O}}:H$

+

molecule

+

molecule

Part 4: Put the molecules together and draw a drop of water, a drop of matter. =

NAME ______________________________ DATE ________________

Atoms Are the Matter

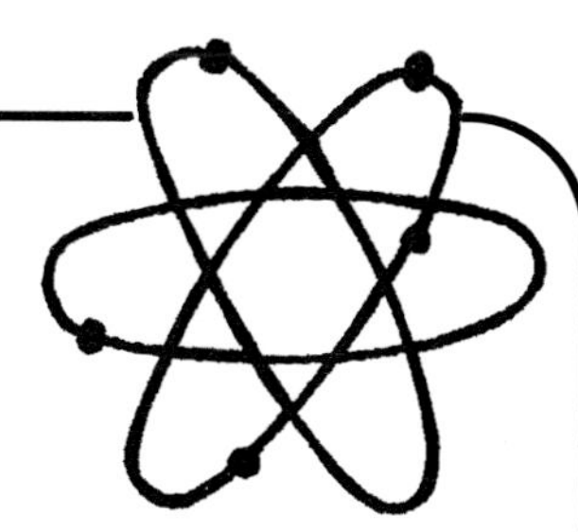

1. **All things are made of atoms.**

2. **Atoms that are the same size and shape as each other are the same type of atom. Atoms that are different from each other are different types of atoms.**

(a) Same	(b) Different

3.(a) **Atoms combine with each other to make molecules.** Molecules make up matter.

3.(b) A picture of Lego water matter.

4. **Sometimes when two different types of molecules get together, they switch atoms around and make different molecules. This is called a chemical reaction.**

(a) I made these molecules	(b) into these molecules

When chemical reactions happen, atoms cannot be created or destroyed.

NAME ______________________________ DATE ________________

For my notebook

The States of Matter

Everything made of atoms and molecules is called matter. Matter comes in three forms called the states of matter. The three states of matter are solid, liquid, and gas. Atoms are the beginning building blocks for all matter. Therefore, all atoms exist as solids, liquids, or gases. Matter, in all three states, takes up space and has mass.

Each state of matter has physical properties that make it special. Think about shape. Solids keep their shape. It would be strange to put a rock in a glass and have it change to the shape of the glass! Liquids and gases do change their shape, though, to match their containers. If you pour tea from a teapot into a teacup, the tea does not stay in the shape of the teapot; it spreads out to take the shape of the cup.

Another physical property that is different from one state of matter to the next is volume. The volume of something is the amount of space it takes up. The bigger something is, the bigger its volume is. For example, your house has a much bigger volume than a mouse. Your house has a bigger mass than a mouse, too.

Solids and liquids have a definite volume. Gases do not. That means that solids and liquids take up the same amount of space even if the size of their containers change. Liquids will change their shape to match the shape of their containers, but they stay the same size. Gases move around to fill their containers. If you poke a hole in a helium-filled balloon, the gas molecules from the balloon will spread throughout the room. The volume of the gas will increase from the size of the balloon to the size of the room because gas is a form of matter that does not have a definite volume.

The States of Matter Lab: Presto-Change-o Water - instructions

CAUTION: This lab uses a heat source and boiling water. Only the parent/instructor should operate the heat source and handle the boiling water.

Materials:
- Lab sheet, pencil
- Ice (about 2 cups)
- Water (about 4 cups)
- Pot for boiling water
- Heat source to boil water
- Three glass containers
- Three sealable baggies
- Rock
- Kitchen scale that measures grams
- Drinking straw (optional)

Part 1

1. Set up the experiment before class. There should be three containers: one with ice cubes, one with water, and one with water to be boiled.
2. At the start of the experiment, boil the water. Make sure it is hot enough to see the steam escape. You might need to heat the water up periodically through the course of the experiment. Have students draw pictures of water in its three states. Then read the following Aloud portion. After you have read this, have students complete Part 1 of the lab sheet before going to Part 2.

Aloud: Matter comes in three forms: solid, liquid, and gas. Water can be in all three states of matter at normal temperatures. Ice is what we call water when it is a solid. The ice melts. Then we have liquid water. If the water is heated, it turns into steam. Steam is what water is called when it is a gas.

Part 2

3. For Part 2 use three baggies: one with a rock in it, one with water in it, and one you have blown air into (a straw can be helpful for blowing air into it). Seal all the baggies (remove as much air as possible from the baggies with a rock and water before sealing). Give these to the students and have them examine how the three types of matter feel and fit in the baggies. Weigh the baggies. The air will probably not show any mass on the scale. Assure students that it is so light that though it is not measurable on a kitchen scale, it does have mass.

Aloud: The three different states of matter have different physical properties. All matter takes up space and has mass. I know that gas doesn't weigh very much, but it does have mass.

4. Have students fill in the first two columns of the Data Table for Part 2.

Aloud: Next put the three baggies in front of you and look at the different properties of the three states of matter. The next two columns on the Table ask about shape. When thinking about shape, ask yourself this: If I pour what is in this baggie into a container with a different shape, such as a teacup, will the shape change? If the shape will not change, then the matter has its own shape. Solids have their own shape. But the liquid will take the shape of the part of the container that it fills, so it takes the shape of the container. Gas takes the shape of the container, too, even if the container is the room you are in.

(continued on the back)

5. Have students fill in the next two columns of the Data Table.

Aloud: The last column asks about volume. The volume of something is the amount of space it takes up. If something has a definite volume, then the amount of space it takes up does not change, even if you move it to a different container. Will the rock take up more space if you put it in a bigger baggie? If not, then the rock has a definite volume. What about the liquid water? The shape of water changes if you move it from one container to another, but does the size of it change? Now look at the baggie with air in it. If you open the baggie what will happen to the air? Actually, it will mix with the air in the room. It will not stay the same size, will it? The size of gases change as the size of the container they are in changes.

6. To demonstrate that liquids have a definite volume, fit the sealed baggie with the water in it in a measuring cup. Then open the baggie and pour the water into the same measuring cup. Notice that the size does not change (even if the shape did) and the water does not try to become as big as the room (like a gas would).

7. Have students fill in the last column on their lab sheets.

Part 3

Aloud: The next part of the lab is a scavenger hunt. Look around inside and/or outside for solids, liquids, and a gas to finish the lab.

8. Give students the scavenger hunt list and a pencil and have them complete the lab. If you have more than one student, you could make it a race.

Answers:

	Takes up space	Has mass (weight)	Keeps its shape	Takes shape of container	Has definite volume
Solid	Yes	Yes	Yes	No	Yes
Liquid	Yes	Yes	No	Yes	Yes
Gas	Yes	Yes	No	Yes	No

Possible answers for scavenger hunt:

Find three solids	Find three liquids	Find one gas
chair	water	air
house	chocolate milk	
pack of cards		

Bonus: helium

Instructor's note:

- There is a fourth state of matter called plasma which includes fire and lightning.

NAME ______________________________ DATE ________________

The States of Matter Lab: Presto-Change-o Water

Part 1: Draw pictures of the water as it looks in the three states of matter.

Ice	Liquid	Gas

Part 2: Let's compare the states of matter.

	Takes up space	Has mass (weight)	Keeps its shape	Takes shape of container	Has definite volume
Solid					
Liquid					
Gas					

Part 3: Scavenger Hunt

Search around your house or school and find matter in its three states.

Find three solids	Find three liquids	Find one gas

Bonus: Can you think of another gas? Hint: you have performed an experiment with it.

NAME ______________________________ DATE ________________

For my notebook

Let's Get to the Point

Have you ever watched a pot of water until it boiled? Maybe you were waiting so you could make macaroni and cheese. There are special names for the points where matter changes from one state into another. The exact point where water goes from very, very, very hot to just starting to boil is called its boiling point. The boiling point of water is the point where water goes from a liquid to a gas. If you go in the other direction from a gas to a liquid, that point has a name, too—it is called the condensation point.

The point where a liquid, like water, goes from a liquid to a solid has a name, too. It is called the freezing point. If you put a tray of ice into the freezer, the exact point where the water goes from as cold as it can be without any ice, to ice beginning to form is the freezing point.

Then there is the melting point. This is the point where a solid turns into a liquid. Some things need a lot more heat than ice does to melt. Have you ever seen a picture of lava? Lava is melted rock. Think about how high the melting point of rock must be!

The points where matter changes from one state into another are measured as temperatures. The temperature these points occur at depends on what's present. For example, the freezing point of water changes if you mix salt into the water. A more dramatic example can be seen when comparing the element nitrogen to that of the element carbon. Remember, these two elements are next door to each other on the periodic table.

Nitrogen–	Melting point:	-346 °F	Carbon–	Melting Point:	6332 °F
	Boiling Point:	-320 °F		Boiling Point:	8721 °F

Let's Get to the Point Lab #1: What Is the Point? - instructions

CAUTION: This lab uses a heat source and boiling water. Only the parent/instructor should operate the heat source and handle the boiling water.

Materials:

- Lab sheet, pencil
- 1 cup Crushed ice cubes
- Clear glass
- 4 cups Water (distilled water is best)
- Heat source for boiling water
- Small pot for boiling water
- One or two science thermometers
- Pot holder
- Science encyclopedia or Internet access (optional)

Aloud: Water freezes, melts, boils, and condenses at specific temperatures. Let's find out what these temperatures are.

Procedure:

1. Put the ice in the glass and set it out to melt. Fill a small pot halfway with water and start it to boil.
2. Just when the ice starts to melt and there is some liquid beginning to form, put the thermometer in the center of the ice water and wait three minutes. Notice that there is ice and melted ice (liquid water) in the cup when the freezing point is measured. Have students write the temperature down on the lab sheet above the words freezing point and melting point.*

LET THE THERMOMETER GET TO ROOM TEMPERATURE BEFORE PUTTING IT INTO THE BOILING WATER. A COLD THERMOMETER PUT INTO HOT WATER CAN SHATTER.

3. Put the thermometer in the boiling water. Hold it in the middle of the water, not touching the sides or bottom of the pot. Wait until the temperature has stopped rising on the thermometer. Notice that as the water starts to boil (boiling point), it also starts to steam and drops of water form (condensation point) on the sides of the pot, or on the lid if you use one. Have students write this temperature above the words *boiling point* and *condensation point*.

Aloud: Did you find that the freezing point and the melting point occur at the same temperature? Because they do. Likewise, the boiling point and the condensation point happen at the same temperature.

Possible Answers:

Freezing point = melting point = 32˚F = 0˚C
Boiling point = condensation point = 212˚F = 100˚C

The point at which a liquid starts to become a solid is called the freezing point.
The point at which a solid starts to turn into a liquid is called the melting point.

The point at which a liquid starts to turn into a gas is called the boiling point.
The point at which a gas starts to turn back into a liquid is called the condensation point.

Instructor's Notes:

- If your tap water has any solids in it (as most tap water does), the temperatures of the points will vary. To get absolute freezing and boiling points, use distilled water and make sure your pot and glass do not have any calcium deposits (chalky white stuff). Temperatures will also vary at extreme elevations.
- *The freezing point is the same as the melting point because it is the exact temperature point where the water in the glass that is ice will remain ice, and the water in the glass that is liquid will remain liquid.

(continued on the back)

NAME ______________________________ DATE ________________

Let's Get to the Point Lab #1: What Is the Point?

Results/Observations: Write the temperatures for each point in the box.

[ ] Freezing Point

[ ] Melting Point

[ ] Boiling Point

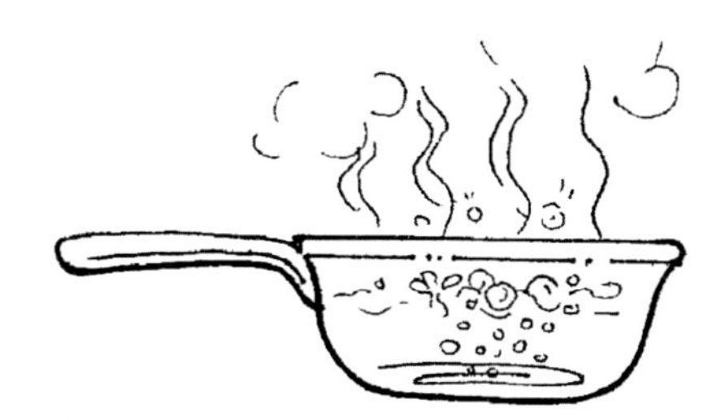

[ ] Condensation Point

The point at which a liquid starts to become a solid is called the

____________________ point.

The point at which a solid starts to turn into a liquid is called the

____________________ point.

The point at which a liquid starts to turn into a gas is called the

____________________ point.

The point which a gas starts to turn back into a liquid is called the

____________________ point.

Let's Get to the Point Lab #2: State of Confusion - instructions

CAUTION: This lab uses boiling water. Only the parent/instructor should handle the boiling water.

Materials:

- Lab sheet, pencil
- Completed lab sheet from The States of Matter Lab: Presto-Change-o Water (p. 279)
- Box of jello
- Cold water and boiling water to make jello
- 1 tablespoon Peanut Butter
- 1 tablespoon Mayonnaise
- Two cups or small containers
- Two spoons
- Bowl
- Refrigerator
- Plate

Aloud: It is easy to tell when water is a solid, liquid, or gas. Some things are easy to categorize, but some things are not. What about peanut butter, mayonnaise, and jello? Are they solids, liquids, or gases? Do they have a definite shape? Are they more like a rock or like water? Do they have a definite volume? Do they behave like a gas? Or more like a solid or liquid? Can something be more than one state of matter at a time? Let's be chemistry detectives and figure this out.

Procedure:
Part 1

1. Complete the hypothesis portion of the lab sheet.
2. Take a spoonful of mayonnaise and peanut butter out of each jar. Examine the spoonfuls and complete the first column on the table by writing yes, no, or maybe.
3. Drop each spoonful into a separate cup. Examine and write *yes*, *no*, or *maybe* in the second and third columns on the table.
4. Compare the table to the completed data chart found on the The States of Matter Lab: Presto-Change-o Water lab sheet.
5. Complete the Results portion.

Possible Answers:
Keeps it Shape - No for both
Takes Shape of Container - No or sort-of for both, depending on how thick the mayo and peanut butter are.
Definite Volume- Yes for both
Notice that mayonnaise and peanut butter have properties of both a solid and a liquid. Neither is a gas.

Instructor's Note:

- Peanut butter and mayonnaise are mixtures. They are both in a state of matter somewhere between a liquid and a solid. Their shape is not completely definite or completely indefinite.

Part 2
Aloud: Jello is fun to make and fun to eat. We are going to make jello today. As we do, pay attention to the state of matter jello is in at each step.

6. Make jello according to the directions on the box. While you are making jello, discuss the different states of matter you are observing. Peek at the jello every hour for three hours while it is setting in the refrigerator to observe it as it changes from a liquid.
7. Record observations on the lab sheet.

(continued on the back)

8. When the jello has "jelled," take it out of the refrigerator and scoop some onto a plate.
9. Complete the lab sheet.

Possible Answers:

Jello comes out of the package as a solid.

It is mixed with a boiling liquid, water. The water is boiling, so there is also gas (vapor) present.

The jello dissolves in the water. At this temperature, you have a liquid.

With every hour that passes with the jello in the refrigerator, the jello will progressively become more like a solid and less like a liquid. Answers can vary between liquid, solid, or both.

When it is ready to eat, jello has sort of a definite shape and a definite volume, so it is a solid—a very jiggly one, though. Some students may answer solid and liquid, which would be correct, too.

Instructor's Note:

- Jello is a colloid. A colloid is a solid suspended in a liquid. It has properties somewhere between a solid and a liquid. So jello, when it has cooled, is a solid (collagen) suspended in a liquid.

NAME ______________________________ DATE ________________

Let's Get to the Point Lab #2: State of Confusion

Part 1 **Hypotheses:**

(Circle your choice. You may circle more than one state.)

I think mayonnaise is a

solid liquid gas

I think peanut butter is a

solid liquid gas

Test: (Write yes, no, or maybe.)	Keeps Its Shape Did it keep the same shape when you scooped it out of the original container?	Takes Container Shape Did it spread out to take the shape of the cup you placed it in?	Definite Volume Did it stay the same size when you placed it in the cup?
Mayonnaise			
Peanut butter			

Results: (Circle your choice. You may circle more than one state.)

Now I think mayonnaise is a solid liquid gas

Now I think peanut butter is a solid liquid gas

Part 2 - Write the correct state of matter: solid, liquid, or gas.

When I poured the jello out of the box, it was a ______________________.

The water I mixed into the jello was a ______________________.

After I stirred the jello and water together, the mix was a ______________________.

After sitting in the refrigerator for 1 hour, the jello mix was a ______________________.

After sitting in the refrigerator for 2 hours, the jello mix was a ______________________.

After sitting in the refrigerator for 3 hours, the jello mix was a ______________________.

When the jello was ready to eat, the state(s) of matter was/were

____.

NAME ______________________________ DATE ________________

For my notebook

Solids Are Dense

When you think of something that is a solid, you think of something with a definite shape. That is unless you took a hammer to the solid. Hammering most solids changes their shape but it takes some force. You would feel it in your arm. That is because in a solid, the molecules are close together, and they do not like to move around much. When you hammer a solid, you are forcing the molecules in the solid to move. These molecules don't like to change position relative to each other; they are not fluid.

If you did hammer a solid, you would be changing the shape of it and sometimes the volume, too. Remember, the volume of something is the amount of space it takes up. Some solids break apart when hammered, like ice cubes. All the pieces of ice would take up the same amount of space, though, if you put them back together. It would be an ice cube puzzle.

What about when you hammer wood? Have you ever noticed that you can dent a piece of wood? When you do this, the wood takes up less space, so you have changed (decreased) its volume. There are still the same amount of molecules and atoms in the wood; they just take up less space. When the hammer forced the same amount of molecules into a smaller space (decreased its volume), the density of the wood increased.

Density is a physical property. It measures the amount of stuff in a given space. The density can be changed by changing either the amount of stuff or the size of the space.

Let's take a closer look at the wood. When the wood is pounded with a hammer, it changes the amount of space the wood takes up. Remember, though, the number of molecules in the wood did not change. That means there is the same number of molecules (stuff) in a smaller space. If the same amount of molecules are in a smaller space, that means the wood is more dense.

The density can also be changed by changing the amount of stuff in a space. Let's do a thought experiment and see how this works. You are having a birthday party. You have invited eleven friends to it. All twelve of you decide to see how many people can fit in your closet. Each of you goes into the closet one at a time. Let's measure the density of people in the closet. Now remember, the closet stays the same size. Each time one more person goes into the closet, the density of people in the closet increases by one. When all twelve of you have managed to get into the closet, you have reached maximum density—no more can fit. Then your mom comes looking for you. She tells all of you to get out of the closet, NOW! As you and your friends exit the closet one by one, the density of people in the closet decreases.

So now that you understand density, let's summarize what makes a solid a solid. In a solid 1) the molecules are close together, and 2) the molecules do not move about much, they vibrate slightly, and they have fixed positions relative to one another. They are not fluid as they are in a liquid or a gas. These characteristics lead to solids having a 3) definite shape for the most part, and a 4) definite volume for the most part. Solids are 5) more dense than liquids with the important exception of water, and 6) more dense than gases.

Solids Are Dense Lab #1:
Some Are Denser than Others - instructions

Materials:

- Lab sheet
- Colored pencils
- Medium-size box or container*
- Several (10 or more) stuffed animals or something else you have 10 of—like oranges, for example

*Choose a size of a container that will allow you to fit all 10 things in it; but when you do, it will be filled to the top.

Procedure:

1. Put one stuffed animal in the box and show that the box's density is not very great because there is a lot of extra room. Explain to students that they are going to increase the density in the box.
2. Slowly increase the density of stuffed animals in the box. Try to envision the amount of stuff, i.e. the number of stuffed animals, in a space increasing without the size of the space changing.
3. Cram the last stuffed animal into the box and explain that the density of stuffed animals in the box is now very great because the box has a lot of matter (stuff) in it and there is not a lot of extra room.
4. Have students demonstrate they understand the concept of density by completing the lab sheet.

Answers:

There should be many circles in the box that is the most dense, fewer circles in the dense box, and the fewest circles in the box that is the least dense.

For More Lab Fun:

If you have a large group, this lab can be done with real kids in a real closet.

NAME ________________________________ DATE ________________

Solids Are Dense Lab #1: Some Are Denser than Others

Fill each box on this page with colored circles. Each circle stands for a molecule. Look at the labels to decide which box gets more molecules and which gets less. Each box should have a different amount.

Least Dense

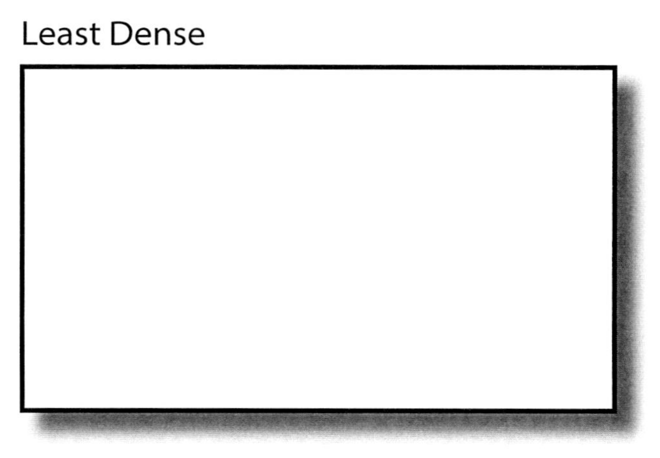

Dense

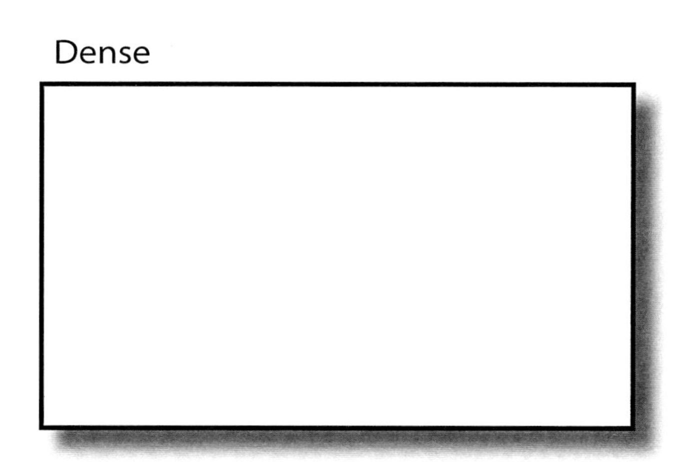

Most Dense

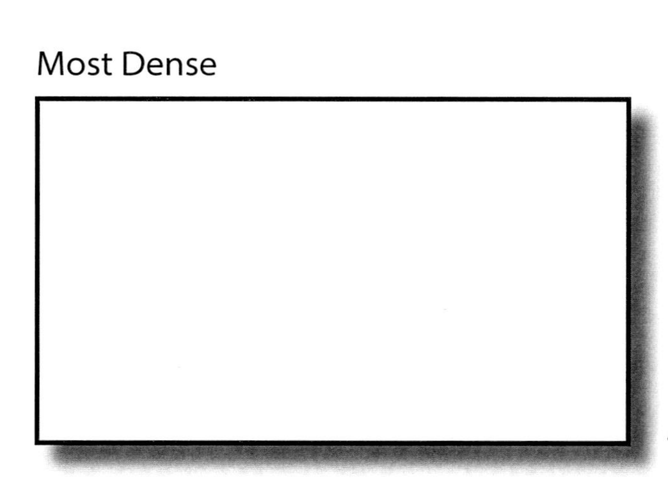

Solids Are Dense Lab #2: The Sinking Tub Boat– instructions

Materials:
- Lab sheet, pencil
- Wash tub, bathtub, or sink filled with water
- One orange
- Small (1- to 2-cup size) plastic container (the "tub boat")
- Marbles (10 to 20, depending on the size of your "tub boat")
- An assortment of eight other solids to drop in the water—plastic and metal toys, stones, sticks, piece of Styrofoam, flower, bar of soap, golf ball, rubber ball, pencil, water proof shoe, etc. Do not use ice (it is the one solid that is less dense than its liquid state—this will be discussed later).

Aloud: Not all solids have the same density. From one type of solid to the next, the density can be different. This happens because from one type of solid to the next, the amount and type of molecules present can be different.

Most solids are more dense than liquids. Liquids are more dense than gases. Because of this, when you mix a gas with a solid, the combination creates a solid whose density might be less than a liquid. Styrofoam is just such a mixture. It is one example of a solid that is less dense than water because of the air trapped in it. The density of an object affects whether it floats or sinks. If something is more dense than water, it will sink in water and if it is less dense than water, it will float in water. For instance, if you drop a rock into water, it will sink right away. However, if you drop an inflated balloon onto water, it will float. That is because the rock is more dense than water, and the inflated balloon is less dense than water. Today you are going to look at how the density of objects affect whether they sink or float.

Procedure:
1. Record the assorted solids you choose on the lab sheet and write predictions about which things will float and which will sink. Students can hold the solids in their hands to feel the mass first, if that helps with predictions. Have students complete the data table as they go through the experiment, so they do not have to remember at the end which floated and which sank.
2. Make a prediction on how many marbles it will take to sink the "tub boat."
3. Fill a sink or tub with enough water to test if items sink or float.
4. Drop the orange in the water. Now peel the orange, drop it in after it is peeled. Record the results.
5. Have students drop the things they collected into the water. Record the results.
6. Put the plastic container (the "tub boat") on the water and watch it float. Now put the marbles into the tub boat one at a time. You might need to stabilize the container so that it doesn't turn on its side and dump all the marbles into the water. Record the results.

Instructor's Note:
- When students are putting marbles into the tub boat, discuss that they are putting more "stuff" into the container with each marble. They are increasing the density of matter in the container.

Possible Answers:
An orange with a peel will float.
An orange without a peel will sink.

Discussion and Conclusion:
1. Peeling the orange changed it from a floater to a sinker because the peel traps air, a gas, between the skin and the flesh, making it less dense. This works in the same way that a life jacket keeps a person afloat.
2. The marbles (a solid) displaced air (a gas) in the container. This made the tub boat more dense. Therefore it sank.

NAME ______________________ DATE ______________

Solids Are Dense Lab #2: The Sinking Tub Boat

Items	**Prediction:** Sinker or Floater	**Observation/Results:** Sinker or Floater
1. Orange with peel		
2. Peeled orange		
3.		
4.		
5.		
6.		
7.		
8.		
9.		
10.		
How many marbles will it take to sink the tub boat?		

Discussion and Conclusion:

1. Did peeling the orange change whether it sank or floated? Why?

2. Why did adding marbles to the tub boat make it sink?

3. Were you surprised by anything that sank or did not sink?

NAME ______________________________ DATE ________________

For my notebook

Molecules Stick Together

Look around you. Think about all the different types of solid matter you can see. All these solids are groups of molecules made of atoms that stick together. Do you remember how small atoms and molecules are, though? They are so small that you could not see just one of them. That means the solid matter you are looking at is made up of millions and billions (and a lot more than that, actually) of molecules stuck together.

You already know that atoms make molecules by sharing electrons. Just like atoms, molecules want to play with other molecules. When molecules want to play, they have to share electrons sort of like atoms do. Their favorite electrons are the ones in their own molecule, but molecules will share back and forth playing together. When this happens and they are all playing together, a solid or a liquid forms. They play together in large groups, too—large enough for you to see and feel them, or large enough, even, to make a house.

The closer that molecules are together, the more they play with each other. The more they play with each other, the less they move around. Molecules that stay close together and do not move around much, are solids. Molecules that still like to play together, but just a little more independently, are liquids. Molecules that do not need other molecules to play form gases.

Think of it this way—let's say you have a neighbor who is a good friend. Now pretend you and your friend are molecules. If you and your friend were at your home playing Monopoly together, you would be sitting close, not moving around much, while you played back and forth. You two would form a solid. The next day you two decide to play catch. When you play catch, you are not sitting close to each other and you are moving around. You still need each other to play, though. On that day, you two would form a liquid. On the following day, you both stay at your own homes and watch the same show on television. In the evening, you meet on the street as you are out walking. You two are walking in different directions. You briefly mention the television show as you pass each other and then you keep walking in separate directions. Today you two are gas molecules.

All this talk about play makes me want to play!

Molecules Stick Together Lab:
A Big Rock Candy Mountain! - instructions

This lab takes about a week to complete, with students recording observations every couple of days.

CAUTION: This lab requires the use of a stove and a very hot sugar solution that will burn if touched. Students should only observe and not handle the hot solution.

Materials:

- Lab sheet, pencil
- 3 cups Sugar
- 1 cup Water
- Saucepan
- Stove top
- Glass jar
- One 6-inch long piece of rough string or yarn
- Wooden spoon
- A clean metal washer or a Lifesaver candy
- Funnel (optional)

Aloud: Sugar is a big molecule, with 12 carbon atoms, 22 hydrogen atoms, and 11 oxygen atoms in it. Sugar molecules like to play with other sugar molecules and make large solid sugar crystals. Today you will take sugar, a solid, and mix it with water, the universal solvent. The sugar is the solute, and it will dissolve in the water, making a sugar-water solution that is a liquid. You will mix so much sugar into the water that the sugar will come back out of the solution, reforming into a solid.

Procedure:

1. Instructor: Pour 3 cups of sugar and 1 cup of water into a saucepan. Heat on medium high, while stirring occasionally, until all the sugar has dissolved.
2. Complete the Before portion of the lab sheet.
3. Instructor: Carefully pour the hot solution into the glass jar. A funnel helps a lot with this.
4. Tie the string to the middle of the pencil. Tie the washer or Lifesaver to the other end of the string. Lay the pencil across the top of the jar so the string end with the washer hangs into the solution but does not touch the bottom of the jar. You can cover this lightly with a sheet of paper to keep dust out.
5. Have students check the jar, as indicated on the lab sheet, for one week and record their results on their lab sheets.
6. On day 7, take out the crystal and dry it on a piece of wax paper or cling wrap.

Instructor's Notes:

- Be VERY CAREFUL—this solution can burn you!
- Rock candy crystals will start forming the first day. The trick is to wait longer than that. The crystals will keep growing for about a week. Try to observe the crystals over the course of a week, without disturbing them. You will have better crystals that way.
- This type of solution is called a super-saturated solution. Crystal formation is a result of the solution being supersaturated and the evaporation of the water.
- When putting the string in the solution, make sure it does not touch the bottom of the jar. If it does, crystals can form on the bottom of the jar and not as well on the string.
- When the rock candy is eaten, remember there is a metal washer (if used) tied to the bottom of the string.
- Food coloring and/or flavor extract (e.g. vanilla, cherry, lemon, peppermint) can be added to the hot sugar-water solution.

(continued on the back)

Possible Answers:
Sugar is a solid.
Water is a liquid.
The sugar-water solution is a liquid.
There should be a progression of drawings over the course of a week showing the rock candy growing.
Rock candy is solid.

Bonus: The candy changed from a liquid to a solid as water evaporated. When moving from a liquid to a solid, the molecules stick together by sharing more electrons, moving closer together, becoming more dense, and not moving around as much.

For More Lab Fun:
Look at the rock candy with a hand lens to get a closer view of the sugar crystals.

Group Activity: With two or more students, pretend you are molecules in a solid, liquid, and gas. To pretend to be a solid, huddle together and play a card game, Dominoes, or read a book aloud. To be a liquid, go outside and throw a ball or Frisbee to each person. Talk about how you cannot get too far away and still play together and that you move a lot more than you did when you were a solid. To be a gas, have everyone walk in all different directions. Walk past each other, then turn around and walk toward each other again. Every time you pass a person, stop and say the name of an element, then keep walking.

NAME ______________________________ DATE ____________________

Molecules Stick Together Lab: A Big Rock Candy Mountain!

Before I make my rock candy:

The sugar is a solid liquid gas.

The water is a solid liquid gas.

The melted sugar-water solution is a solid liquid gas.

Drawings of My Rock Candy

One hour after starting	One day after starting	Two days after starting
Four days after starting	**Six days after starting**	**When I took it out and ate it!**

The completed rock candy is a solid liquid gas.

Bonus: Describe how the rock candy changed from one state of matter to another and what happened to the molecules as this change was happening.

____

____

____

NAME ______________________ DATE ______________

For my notebook

What Makes a Liquid a Liquid?

What would happen if you picked up a hammer and pounded it into a liquid? Would you feel the force of it in your arm or would you just get wet? You would feel some resistance when your arm hit the liquid as the liquid slowed your arm down. It does not make sense to hammer a liquid, but if you did, you would change its shape for a short time as it rippled. Also you would change the volume, the amount of space it took up, but only for as long as it took you to remove your arm and the hammer from it. That is because in a liquid, the molecules can move around. If you try to hammer a liquid, the molecules move to make space for your arm and the hammer. When you take your arm out again, the liquid moves back to the original shape, which is the same shape as the container holding the liquid. The molecules in a liquid like to play with each other, but they like to move around, too.

So now you probably know enough about liquids to know that the characteristics of a liquid are 1) the molecules are close together but not as close as in a solid, 2) the molecules in a liquid move around but not as much as in a gas, and 3) the molecules in a liquid are <u>fluid</u>. These characteristics lead to liquids having 4) a definite volume but 5) not a definite shape. So liquids are 6) less dense than solids, with the important exception of water, and 7) more dense than gases. But you already knew that!

What Makes a Liquid a Liquid? Lab: Liquids Are Dense, Too - instructions

Materials:

- Lab sheet, pencil
- ¼ cup Corn syrup
- ¼ cup Liquid cooking oil - olive oil, vegetable oil, or any kind you have on hand
- ¼ cup Water
- Clear container that holds 2 or more cups of water
- Measuring cup
- Food coloring

Aloud: Just like solids, some liquids are more dense than others. If you put two liquids of different densities in the same container, the more dense liquid will sink to the bottom and the less dense liquid will float on top.

Procedure:

1. Have students fill in the hypothesis section of the lab sheet.
2. Mix food coloring with the water. This will help to distinguish between the different liquids.
3. Tilt the empty glass on its side and carefully pour in the corn syrup, then add the liquid oil, and finally add the colored water into the glass.
4. Gently set the glass back down and wait for the layers to stabilize. How fast this happens depends on how much the liquids were shaken while pouring and setting the glass down.
5. Have students color and label the glass on the lab sheet and complete the rest of the questions.

Possible Answers:

The liquid that is the least dense = oil
The liquid with the density in the middle = water
The liquid that is the most dense = corn syrup

Bonus: The molecules in a more dense liquid are packed closer together in a smaller space than liquids that are less dense.

NAME ______________________________ DATE ________________

What Makes a Liquid a Liquid? Lab: Liquids Are Dense, Too

Hypotheses: Which liquid do you think is more dense?

Label your predictions below: water, oil, and corn syrup

The liquid that is least dense =
The liquid with the density in the middle =
The liquid that is the most dense =

Results/Observations: Color and label the layers in the glass

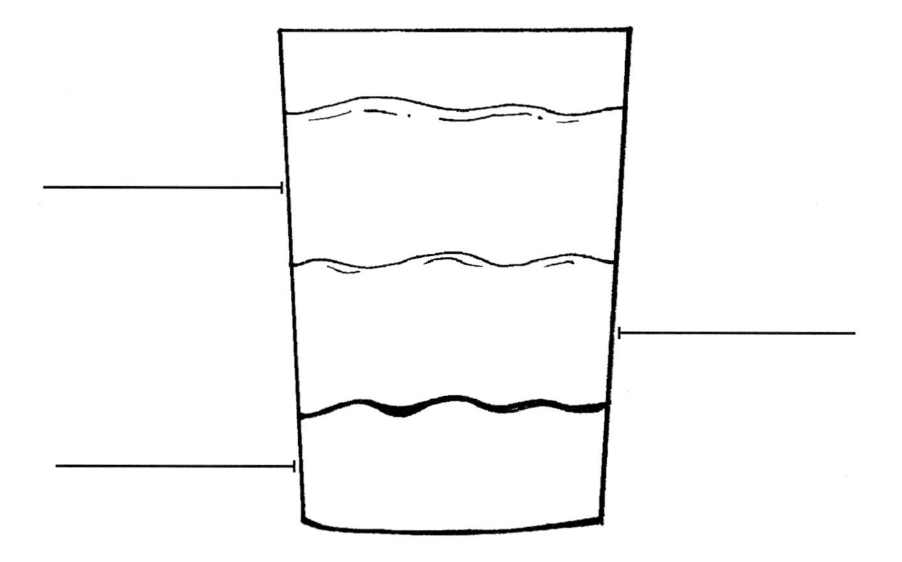

Conclusion: How did the liquids layer?

The liquid that is least dense =
The liquid with the density in the middle =
The liquid that is the most dense =

Bonus: Remembering what you have learned about density, what makes one liquid more dense than another?

Drawing the States of Matter - instructions

Materials:

- Lab sheet, pencil
- Board or sheet of paper for you, the instructor, to write on
- Slime, if you still have it from The Slime That Ate Slovenia lab (optional)

Procedure:

1. Have students draw a table, milk, and air in the first column on the lab sheet. Students should fill in their lab sheets before anything is read to them. Some students will be creative when they are asked to draw milk and air, and some will be stumped. Let students puzzle through this. If they ask for it, help them with whichever path this mental exercise takes them. The goal is to reinforce the concepts of definite shape and definite volume. Remember that only solids have a definite shape, and solids and liquids have a definite volume (the amount of space something takes up). If they draw a container in which to hold the milk and the air, that is okay, but ask them to explain why. Some very resourceful students might draw the molecules in milk and air.
2. You might want to write down the phrases "definite shape" and "definite volume" and review their meanings.

Aloud: The table should have been easy to draw. Did you find it as easy to draw milk or air? Because they do not have a definite shape, they are not as easy to draw unless a solid, like a glass or a balloon, is surrounding them. It would be a lot easier if you were asked to draw a glass of milk and a balloon filled with air.

3. Now is a good time to discuss the specifics of what students drew. Ask questions about their thought process. Make sure they really understand the concept of definite shape.

Aloud: Next to your other drawings, draw half a table, a glass half filled with milk, and a balloon that is half filled with air and tied.

4. Wait for students to complete this part of the exercise in the second column on the worksheet.
5. In the next Aloud section, wait for students to answer each question before going on to the next question.

Aloud: Does the milk go to the top of the glass? Or does the milk only go halfway up the glass? Are there air molecules everywhere in the balloon? Or did you draw the balloon with only half of it with air molecules in it? Is the table half the size as it was before? Because solids and liquids have a definite volume, there is now only half a table and half a glass of milk. The balloon, on the other hand, would be smaller if you decrease the amount of air, the air molecules will still fill the entire balloon. This is because solids and liquids have a definite volume, but gases do not.

6. Now is a good time to discuss the specifics of what students drew. Ask questions about their thought process. Make sure they really understand the concept of definite volume.

Possible Answers:

Bonus: What is slime? Slime takes the shape of its container like a liquid. Slime can be picked up in your hand without falling apart, which is not a property of a liquid. Slime is a polymer, as is jello. Polymers are long chain molecules that stretch and bend.

NAME ______________________________ DATE ________________

Drawing the States of Matter

Table 1	Table 2
Milk 1	Milk 2
Air 1	Air 2

Bonus: What is slime—a solid, liquid, or gas?

NAME ________________________________ DATE ____________________

For my notebook

The Friendship of Oxygen and Hydrogen

Water is so common and such a part of everyday life that most of us take it for granted. Water is actually a very special molecule. Water is the most common molecule on Earth. Water is the only substance found in all three states of matter—solid, liquid, and gas, at normal temperatures. About 70% of the Earth's surface is covered with water in either its liquid or solid forms. There is also water in the air. All living things have water in them. The human body is about 65% water. In fact, a human can live only eight to ten days without water.

For almost all substances, the solid form is MORE dense than the liquid form. One unusual property of water is that it is LESS dense as a solid than it is as a liquid. That is why ice cubes float on water. This also means that ice takes up more space than the same amount of liquid water. Maybe water takes up less space as a liquid than as a solid because it is more fun to play together when you are not freezing cold.

This property of water is very important to life in watery places. When the temperature drops and ice starts to form in lakes and seas, ice forms on the top because it is less dense than liquid water. This is important because the ice acts as an insulator and keeps the water underneath from freezing. It insulates the lakes and seas the way your home insulates you and your family. If ice were more dense than liquid water, lakes and seas would freeze from their bottoms all the way up. This would be a problem for animals, like fish, which have to have liquid water to live.

Do you remember that water is a molecule made of hydrogen and oxygen? Hydrogen and oxygen are such good friends that they always want to play together even if they belong to other molecules. The oxygen and hydrogen of water molecules are always playing with the oxygen and hydrogen of other water molecules. When they are playing, they do not like to get too far apart from each other. This playing has a special name; it is called <u>hydrogen bonding</u>. Hydrogen bonds are what hold water molecules together. It is lucky for us that hydrogen and oxygen like each other so much.

The Friendship of Oxygen and Hydrogen Lab #1: Smart Ice - instructions

This lab takes about eight hours or overnight to observe the final results.

Materials:

- Lab sheet, pencil
- Freezer
- Water
- Blue food coloring
- Two empty small plastic soda or water bottles that are the same size, one with its cap
- Pan or cup that is large enough to loosely fit under one of the bottles
- Blue-colored pencil or crayon

Aloud: **Ice is less dense than liquid water. Remember that density measures the amount of "stuff," in this case water molecules, in a certain amount of space. So if frozen water (ice) is less dense than liquid water, then in the same amount of space there are less molecules of ice than liquid water. That means that one measured cup of liquid water has more water molecules in it than one measured cup of ice. If you were to let the cup of ice melt, the result would be less than 1 cup of liquid water.**

As you have already learned, though, if you start out with 2 equal cups of water and freeze one of them, they have the same mass, or weight. That is because the number of molecules does not change—what has changed is the amount of space taken up by these molecules. In other words, the size changes, but the mass does not. Let's experiment and find out what happens to the same amount of water molecules when they go from a liquid to a solid, and become less dense.

Procedure:

1. Put a couple drops of blue food coloring into each bottle. (The food coloring is added only so students can better observe and draw the water.)
2. Fill both bottles halfway with water and shake gently to mix the food coloring. Then fill the bottles the rest of the way with water as full as they can get. These are the same-size bottles, so the same amount of water should fill both of them. Try to make sure you do not leave an air space at the top. Screw the cap onto one of the bottles. The reason for screwing the cap on is so no water evaporates from the bottle.
3. Complete the hypothesis portion on the lab sheet.
4. Carefully put the bottle, the one without the cap, in a pan or cup and place it in the freezer. The pan is to catch the water as it expands. It will spill on the floor of your freezer if you do not do this.
5. Let the bottles sit overnight.
6. The next day, record your observations.

Possible Answers:

Hypotheses:

Initial drawings will vary.

The square representing a view of the water should have more circles in it than the square representing a view of the ice.

Observations:

The water in the bottle that was frozen expanded out of the bottle when it froze.

The water in the bottle that sat at room temperature remained at the same volume.

(continued on the back)

Discussion and Conclusion:
When water molecules go from a liquid state to a solid state, they take up more space. Ice water, in the solid state, is less dense than water in the liquid state. In order to have a decrease in the density, the water molecules must take up more space. Therefore, the water expands out of its container—the plastic bottle.

NAME ______________________ DATE ____________

The Friendship of Oxygen and Hydrogen Lab #1: Smart Ice

Hypotheses: For each of the bottles, draw what you think will happen to the water overnight.

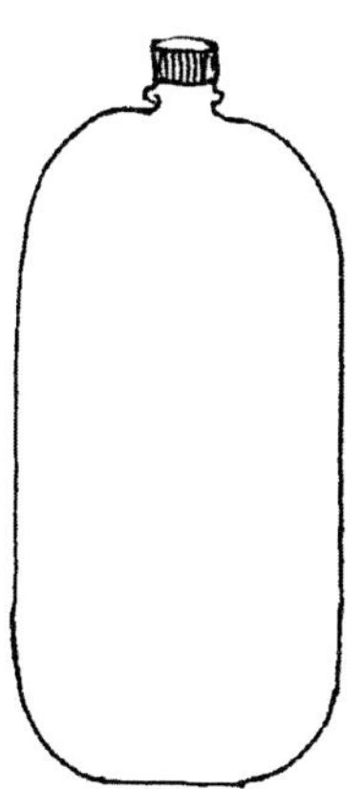

The two squares below are meant to represent microscopic pictures of a small amount of liquid water and ice. Using circles to represent the water molecules, draw a picture showing the change in density that you expect, when going from liquid water to ice.

liquid water

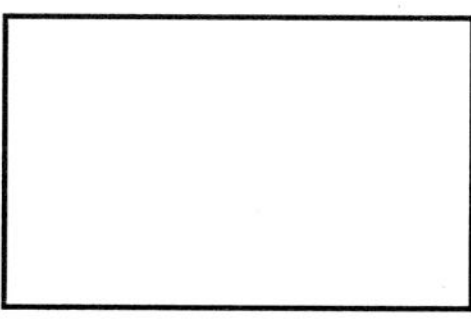

ice

Observations: For each of the bottles, draw what did happen to the water overnight.

Discussion and Conclusion: Based on the results of this experiment, what do you think happens to water molecules when they go from the liquid state to the solid state?

The Friendship of Oxygen and Hydrogen Lab #2:
The Breakup - instructions

Materials:

- Lab sheet, pencil
- Water
- Clear glass tumbler
- Two pieces of insulated copper wire, about 6 inches each
- 9-volt battery
- ½ teaspoon Salt
- Spoon for stirring
- Scissors or knife

Aloud: Have you ever broken up a friendship? If not, today is the day. The experiment today is the electrolysis of water. Water molecules are made from two gases—hydrogen and oxygen. In the electrolysis of water, liquid water molecules are split into the gases hydrogen and oxygen. That means you will be turning a liquid into a gas. Salt is added to speed the process up. You will see lots of hydrogen bubbling to the surface. Oxygen is much slower to leave all of its friends.

Procedure:

1. Fill the tumbler with water to about 1 inch from the top of the glass. Add the salt and stir.
2. INSTRUCTOR ONLY: Use scissors or a knife to carefully strip about 1 inch of the insulation off both ends of both pieces of wire.
3. Wrap one piece of wire around each of the batteries' terminals.
4. Have students draw the setup before putting the wires in the water. Assist them in labeling the parts of the setup.
5. Put the loose ends of the wire into the water. Do not let the ends of the wire touch each other where the insulation has been stripped away.
6. Have students finish their drawings by showing what is happening in the water. Assist them in labeling what is happening in the water (hydrogen and oxygen bubbling on the wires and out of the water).

Instructor's Notes:

- Salt is added to the water to speed up the electrolysis. When salt is used like this, it is called an electrolyte.
- Bubbles will form on the ends of the wire that are in the water. On one piece of the wire, the bubbles will bubble up to the surface. This is hydrogen gas coming off the cathode. The cathode is the negatively charged electrode. The other wire will have bubbles form slowly on it. This is the anode, the positively charged electrode, and this is the wire where the oxygen is forming. The oxygen is not as likely to bubble up to the surface.
- The gas coming from the anode may be chlorine gas (if you have a lot of chlorine in your water) and not oxygen gas. But you will not notice a difference.

NAME ______________________________ DATE ____________________

The Friendship of Oxygen and Hydrogen Lab #2: The Breakup

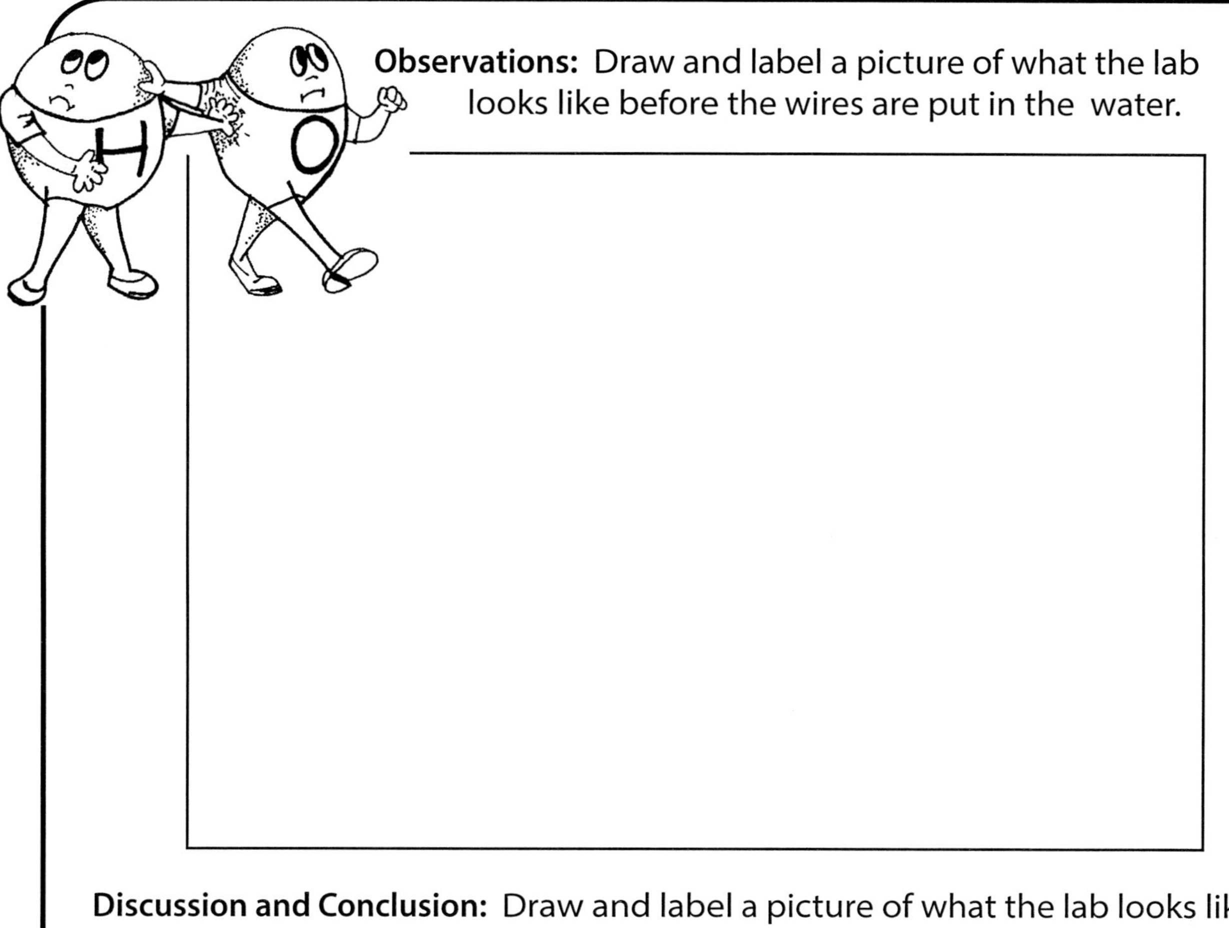

Observations: Draw and label a picture of what the lab looks like before the wires are put in the water.

Discussion and Conclusion: Draw and label a picture of what the lab looks like after the wires are put in the water.

NAME ________________________ DATE ______________

For my notebook

What Makes a Gas a Gas?

Hammering a gas, such as air, would be downright dangerous! If you were not careful, you might hit yourself. Gas molecules move apart easily to let you through. A gas can exert a strong force, though. Have you ever been out in a strong wind? Wind is moving air, and wind can be very powerful.

When you hammer a piece of wood, you do not feel the air as your arm moves through it, do you? You are not forcing the molecules apart from one another when you move through a gas, as you would have to with a solid or a liquid. The molecules in a gas have more space between them than those in a solid or a liquid. They are very fluid. When you move through them, they easily move out of your way without you feeling it. As you walk around and even when you are holding still, gas molecules from the air are bouncing off you all the time. They are so small you don't feel them. It would be really strange if the molecules that make the walls of your house or those from a glass of milk started floating around and bouncing off you, wouldn't it?

So now you know that in a gas 1) the molecules have a lot of space between them, and 2) the molecules move very fast and mix freely because they are very fluid. Because of these characteristics, gases 3) do not have a definite volume, and 4) do not have a definite shape. Gases are 5) less dense than solids, and 6) less dense than liquids. You sure do know a lot about gases!

What Makes a Gas a Gas? Lab #1: Bubbles – instructions

Materials:

- Lab Sheet, pencil
- Unopened can of soda (any kind, including club soda or natural soda)
- Clear glass

Aloud: Soda is a mix of liquid and gas. To give soda its fizz, carbon dioxide, a gas, is dissolved into a liquid. If the bubbles float to the top, what does that tell you about the density of gases compared to that of liquids?

Procedure:

1. Complete the hypothesis portion of the lab sheet.
2. Open the can of soda and pour it into the glass. Observe what happens to the carbon dioxide gas dissolved in the liquid.
3. Record your observations. Complete your lab sheet.

Instructor's notes:

- This is a very simple lab. Its sole purpose is to demonstrate that gases are less dense than liquids.
- This lab is so short that it could be paired with the next lab, Popping Corn.

Possible Answers:

Hypothesis:

The correct answers are C and A. Answers may vary because this is the hypothesis, which is a best guess.

Results/Observation:

The bubbles float to the top and escape into the air.

Discussion and Conclusion:

The bubbles float to the top and escape into the air because the gas, carbon dioxide, is less dense than the liquid part of the soda.

NAME ______________________________ DATE ________________

What Makes a Gas a Gas? Lab #1: Bubbles

Hypotheses: When the soda is poured into the glass, I think the carbon dioxide gas in the soda will (circle the correct answer)

A. sink to the bottom of the glass.

B. spell my name in bubbles.

C. float to the top of the soda and escape into the air.

It will do this because carbon dioxide gas (circle the correct answer)

A. is less dense than the liquid.

B. is more dense than the liquid.

C. likes my name.

Results

Observation: What did happen to the bubbles?

____

____

____

____

Discussion and Conclusion: Can you explain why this happened?

____

____

____

____

What Makes a Gas a Gas? Lab #2: Popping Corn – instructions

CAUTION: This lab requires the use of a stove and hot oil that will burn. This lab also involves cutting with a knife. Students should only observe and not handle the hot oil or use the knife.

Materials:
- Lab sheet, pencil
- Popping corn (you examine a kernel of the corn before popping it; for this reason microwave popcorn does not work, unless you can get a kernel from another source)
- Pan to pop popcorn
- Knife to cut a kernel of popcorn lengthwise
- Oil for popping corn
- Heat source for popping corn—either a stove top or a popcorn popper

Aloud: Liquids are more dense than gases. That means that if you heat liquid molecules until they turn into gas molecules, the gas molecules will take up more space. There will still be the same number of molecules, but they will take up more space. Liquid molecules like to play a lot closer together than gas molecules. Gas molecules need their space.

Today you are going to pop popcorn. Did you know that an unpopped kernel of popcorn has liquid water in it? What do you think happens to the liquid water in popcorn when it is heated? Let's pop some and find out.

Procedure:
1. Complete the hypothesis portion of the lab sheet.
2. Instructor: Take a kernel of popcorn and cut it lengthwise. It is tricky to cut, so students should not do this themselves. Talk about what you see. Take the time now to have students draw a diagram in the observation section. Even if there is no visible water, it is there. The expansion of the water when it goes from a liquid to a gas is what makes popcorn pop.
3. Pop the popcorn, following the directions on the popcorn container. Discuss what is happening in the kernels to make them pop.
4. Record your observations. Complete your lab sheet.

Possible Answers:

Hypothesis:
The correct answer is B. Answers may vary because this is the hypothesis, which is a best guess.

Discussion and Conclusion:
The corn popped because when heated, the water inside of the kernels turned from a liquid to a gas. The gas had to escape, because gases need more room than liquids. The water gas escaped by popping the kernel open.

NAME ______________________________ DATE ________________

What Makes a Gas a Gas? Lab #2: Popping Corn

Hypothesis: What do you think happens to the water in a kernel of popcorn to make it pop? (Circle your answer.)

A. It turns into a solid.

B. It gets hot enough to turn into gas and pops the kernel as it escapes.

C. Nothing happens—the kernel just pops.

Results/Observation:
Draw a picture of the popcorn kernel when you cut it open.

Draw a picture of a piece of the popped corn.

Discussion and Conclusion: In your own words, what do you think happened to the popcorn to make it pop?

____

____

____

____

What Makes a Gas a Gas? Activity: Let's Go Fly a Kite! - instructions

Materials:

- Kite
- A good windy day for flying a kite

There is no lab sheet for this lab.

Aloud: **A kite is a solid. It is also heavier than air. How can a gas, such as air, hold up a solid, like a kite? Airplanes and rockets have engines to help push them through the air, but what about a kite? When you are trying to get a kite up in the air, you run with the kite. When you do this, the kite pushes down on the air. The air pushes back up on the kite. The air pushing up on the kite lifts it into the sky and allows the kite to fly. That is how a gas can hold up a solid.**

Procedure:

- Go outside and fly a kite.
- If you do not have a kite and there is wind or a breeze, stand there and discuss the fact that the wind is made from air. Feel the air molecules (gases) pushing against you. If there is no wind, twirl around. The resistance you feel when twirling is also caused from air.

NAME ______________________________ DATE ________________

For my notebook

The Air You Breathe

You can only live eight to ten days without water, which is not a very long time. Right now, hold your breath as long as you can. Holding your breath makes the eight- to ten-day wait for water seem like a long time, doesn't it? You have just proven that you can go longer without water than you can without air. What is air and why is it so important to our bodies that a minute without a breath seems like a really long time? The oxygen from the air you breathe and the food you eat gives you energy. In fact, people, animals, and plants cycle oxygen back and forth to each other.

As you might have learned in earth science, the air, which is made from gas molecules, forms the atmosphere. The atmosphere is important to all life. Without air, Earth would not have an atmosphere and without an atmosphere, we would not have Earth as we know it. The gases in Earth's atmosphere act as a protective blanket around Earth. During the day, the atmosphere keeps us from being burned by the intense heat of the sun. During the night, the atmosphere keeps us from being frozen by extremely low temperatures.

We spend our entire life surrounded by a gas, but most people would find it easier to describe liquids and solids. That is because the gases that surround us are invisible to us. Air molecules are small even for molecules. Each nitrogen molecule in the air has only two nitrogen atoms in it and each oxygen molecule has only two oxygen atoms in it. Air is 78% nitrogen, N_2, and 21% oxygen, O_2. That means if you had 100 air molecules in your hand, 78 of them would be nitrogen and 21 of them would be oxygen. That would account for 99 of the 100 molecules. The other "molecule" would most likely be an argon atom, Ar. When you look across air, you don't see these molecules, though. Remember how small molecules are? Gas molecules don't like to hang out together, either. Gases have much more space between their molecules than liquids and solids do. The air around you is full of them but you don't see them. Air might be invisible, but we sure are lucky to have it.

The Air You Breathe Lab #1: Air Takes up Space – instructions

Materials:

- Lab sheet, pencil
- Balloon
- Empty 2-liter soda bottle

Aloud: Air does take up space. But it is hard to show that air takes up space, because when you move through air, it is easy for the air molecules to move out of your way. Today you will see what happens to air when the molecules do not have anywhere to go.

Procedure:

1. Complete the hypothesis portion of the lab sheet.
2. Have your student blow up the balloon and then let the air out.
3. Put the balloon into the soda bottle and wrap the end of it around the lip of the bottle. Make sure the balloon is wrapped completely over the lip of the bottle so no air from inside the bottle can escape. Holding the mouth of the bottle to his lips, have your student try to blow up the balloon. Let him work at it until he is convinced that it cannot be done.
4. Record your observations. Complete your lab sheet.

Instructor's Note:

- Air is 99% nitrogen and oxygen. The other 1% is made up of a mixture of carbon dioxide, water vapor, argon, and ozone, primarily. These molecules might be at a low concentration, but they have an important influence on Earth's climate. There are other molecules in the atmosphere as well. What these are depends on where you live.

Possible Answers:

Results:

Outside the bottle: The balloon will blow up normally.
Inside the bottle: The balloon will not blow up, even to fill the inside of the bottle.

Discussion:

When the balloon was blown up outside the bottle, it inflated normally, because the air in the room has plenty of space to move out of the way of the balloon.

When you tried to blow the balloon up inside the bottle, it could not be inflated. This is because the air inside of the bottle could not move out of the way of the expanding balloon.

NAME ________________________________ DATE ____________________

The Air You Breathe Lab #1: Air Takes up Space

Hypotheses: Draw two pictures showing how you think your balloon will look when you try to blow it up.

Balloon outside the bottle

Balloon inside the bottle

Results/Observations:

When I blew the balloon up outside the bottle, it looked like this:

When I blew the balloon up inside the bottle, it looked like this:

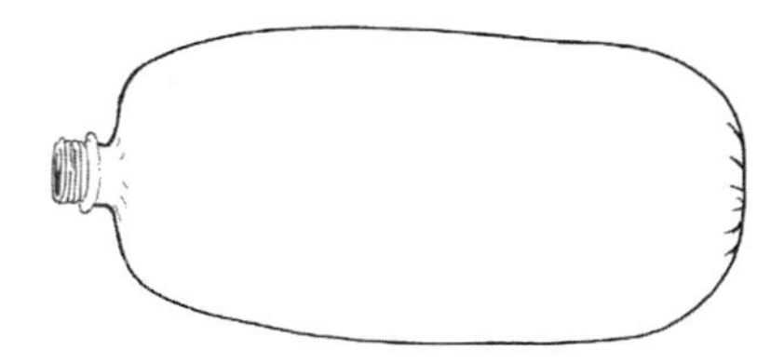

Discussion/Conclusion:

When you blew the balloon up outside the bottle, what happened and why?

When you blew the balloon up inside the bottle, what happened and why?

The Air You Breathe Lab #2: Why Do Boats Float?– instructions

Materials:
- Lab sheet, pencil
- Empty glass jar with a lid
- Tub or sink full of water
- Optional: Movie or documentary about the *Titanic*

Aloud: Why do steel boats float? A big steel boat is heavy. Boats are made of solids, which are usually more dense than liquids. Boats are not just solids, though. They are solids filled with air, which is a gas. This solid-plus-air combination affects the density of a big ship. The air makes the boat less dense, so it can float. The ability of things to float is called buoyancy.

Submarines float and sink. They can control their buoyancy. They do this by changing their density. When a submarine wants to go underwater, it takes on water. When the submarine takes on water, it lets out air. The water is more dense than air. This makes the submarine more dense and allows it to sink. Submarines can control how far they sink by controlling how much water they exchange for air. When a submarine wants to come back up to the surface, the water is pumped back out.

If a boat fills with water, it will sink. Do you know the story of the *Titanic*? On April 14, 1912, a big ship called the *Titanic* collided with an iceberg. The iceberg ripped through part of the ship that was underwater. The air in the ship bubbled out, and water took its place. When this happened, the *Titanic* became too dense to float and it sank.

Glass is more dense than water. If you put a glass jar filled with air into water, how will the air affect the density of the glass jar? Will the glass jar sink or float? What if the glass jar is filled up with water? Will it sink or float then?

Procedure:
1. Complete the hypothesis portion of the lab report.
2. Fill a tub or sink up with water. Put the jar with the lid on tightly into the water on its side. Observe the jar floating. Gently push the jar down with your hand and let it spring back up. Be careful not to push it too far down or it could come shooting out of the water and break.
3. When you are done observing this, remove the jar from the water and take the lid off the jar.
4. Put the jar back on its side into the water.
5. The jar should start to take on water. If it does not, you might need to get it started by pushing it down until water starts to go into the mouth of the jar.
6. Let the jar fill up with water until it sinks.
7. Record observations and complete the lab sheet.

Instructor's Note:
- If you want to have more fun with this, name your "ship." Then pretend that you are an iceberg and when you remove the lid, you have punctured your ship.

(continued on the back)

Possible Answers:

Hypothesis:
Answers will vary.

Observations:
The jar will float when the lid is on.

With the lid off, the jar will float initially, until it takes on enough water, then it will sink.

Discussion/Conclusion:
With the lid on, the gas that is in the jar makes the jar buoyant enough to float. This is despite the fact that the glass is more dense than the water.

With the lid off, the jar takes on water. Once the jar fills up with water, the fact that the glass jar is more dense than the water makes the jar sink.

For More Lab Fun:
Watch a documentary movie about the *Titanic*.

NAME ______________________________ DATE ________________

The Air You Breathe Lab #2: Why Do Boats Float?

Hypothesis:
Do you think your jar will float? Give your reason, why or why not.

Observations:
Draw pictures of what happened to the jar.

Lid on

Lid off

Discussion/Conclusion:
In your own words, explain what happened in this experiment.

What's the Matter? - Crossword Vocabulary Review

Across

1. The amount of space something takes up.
4. How many different states of matter are there?
7. The point where a solid becomes a liquid. (Two words)
8. Has a definite volume but not a definite shape.
9. Does not have definite shape or volume.
11. The point where a gas becomes a liquid. (Two words)

Down

2. Anything that has weight and takes up space.
3. The point where a liquid becomes a solid. (Two words)
5. The point where a liquid becomes a gas. (Two words)
6. The amount of stuff in a given space.
10. Has a definite shape and a definite volume.

Unit 7

Reactions in Action

Chemical Reactions Make the World Go Around

(Sung to the tune of "Money Makes the World Go Around")

Chemical reactions make the world go around
The world go around
The world go around
Chemical reactions make the world go around, just like this . . .
First, you have a hydrogen over here and oxygen over there and a hydrogen over here
Next, you bring them all really close and then you have water
Water, Water
Water, Water
Water, Water
Water

Chemical reactions make the world go around
The world go around
The world go around
Chemical reactions make the world go around, just like this . . .
First, you take a log from the pile, a log from the pile, a log from the pile
Next, you bring a match really close and light the logs on fire
Fire, Fire
Fire, Fire
Fire, Fire
Fire

Chemical reactions make the world go around
The world go around
The world go around
Chemical reactions make the world go around, just like this . . .
First you mix the butter and the sugar with the eggs and the flour.
Next you bake the batch in the oven and then you have cookies
Cookies, Cookies
Cookies, Cookies
Cookies, Cookies
Cookies

Chemical reactions make the world go around
The world go around
The world go around
Chemical reactions make the world go around
The world go around.

NAME ______________________________ DATE ________________

For my notebook

Changes

Are you the type of person who likes to watch things explode? Or maybe you are more the type who takes real pleasure out of mixing flour, eggs, butter, and sugar to make delicious cookies. Then there is the nice, warm feel of a cozy fire on a cold day. All of these are examples of chemical reactions.

A chemical reaction is the combination, separation, or rearrangement of atoms. What this means is, the molecules at the end of the reaction are not the same as the molecules at the start of the reaction. But during a chemical reaction, atoms are not created or destroyed. They just change whom they are playing with.

To understand chemical reactions, first you have to know the difference between physical change and chemical change. A physical change doesn't change what something is. It might look very different, but it is the same thing. For example, ice, water, and steam are all a form of water. When you freeze water, or boil it and turn it into steam, the molecules are still water molecules, H_2O. That's because the change from one state of matter to another is an example of a physical change. If you rip a piece of paper into many pieces or squash a can, these are also examples of a physical change. The shape of the paper and the can have been changed, but they are still made up of the same types of molecules. Another example of a physical change would be a solution of sugar and water. This is a physical change because the sugar can be separated from the water. All you need to do is let all the water evaporate away or insert a rough string.

In a chemical change, new and different molecules are made. You do not have the same molecules at the end that you had at the beginning. If instead of ripping the piece of paper, you burned it, that would be a chemical change. Once burned, you no longer have paper. Instead, there is ash, smoke, and heat. If you used sugar to make a cake, this is an example of a chemical change. Can you imagine getting all the sugar molecules back from the cake again? Neither can I, nor would I want to! If a chemical reaction happens then there has been a chemical change.

Changes Worksheet: Physical or Chemical? - instructions

Materials:

- Lab sheet, pencil
- The text—to look up the labs you performed to help answer some of the questions.

Procedure:

The worksheet has a list of events that involve making changes to matter. All the events are examples of either physical or chemical change. Several of the examples have been taken from labs performed this year. If students need help deciding whether a change is physical or chemical, you can refer back to the lab. The lab names and page numbers are listed in parentheses below.

Aloud: To determine if there has been a physical or chemical change, ask yourself if the matter looks the same after the change. The shape may have changed, but does it look like you have the same material? If the answer is yes, there has been a physical change. If the answer is no, then ask yourself if you can change it back. If you can change it back, such as ice to water, that is usually a physical change as well. If you cannot change it back and/or it doesn't look the same, then there has most likely been a chemical change. Bubbles forming is also usually a sign of a chemical change. You have performed lots of experiments so far this year. Some of those experiments have been examples of chemical changes and some of the experiments have been examples of physical changes. Let's see if you can figure out which ones are which.

Answers and explanations:

1. Physical – The piece of wood has been covered with paint, but it is still the same piece of wood. The wood can be changed back to its original appearance by sanding it with sandpaper.
2. Chemical –Through a chemical reaction, lactic acid is produced. This process cannot be reversed.
3. Chemical – This is an example of a type of chemical reaction called a combustion reaction. After the match has been lit, you no longer have what you started with.
4. Physical (Periodic Play Dough, p. 111) - You started with play dough, and after making the sculpture you still have play dough. FYI, making the play dough is also a physical change. Even though they would be nearly impossible to separate, the ingredients (peanut butter, milk, and honey molecules) are unchanged when mixed together.
5. Physical (Dancing Drops, p. 205) - If you evaporate all the water from the container, food coloring residue will be left.
6. Physical (Telling Things Apart, p. 27) - There is no chemical reaction; they can be separated from each other.
7. Chemical (Telling Things Apart, p. 27) - One of the products from this reaction is carbon dioxide bubbles forming.
8. Chemical (Eating Hockey Pucks, p. 129) - During baking, chemical changes occur. One such change is the creation of carbon dioxide from baking powder.
9. Chemical (My Favorite Element, p. 77) - The citric acid in the lemon juice breaks down the cellulose in the paper and turns the cellulose into sugars. The carbon in the sugars darken when exposed to heat.
10. Physical (A Big Rock Candy Mountain!, p. 303) - You can reheat the rock candy in the solution. The rock candy will go back into solution and you will have what you started with.
11. Chemical (The Breakup, p. 323) - This is an example of a type of chemical reaction called a decomposition reaction. You started with 2 H_20, applied an electrical current to it, and the water decomposed into 2 H_2 and O_2.
12. Physical – Melt the grape juice and you will have what you started with.

(continued on the back)

13. Chemical (The Slime That Ate Slovenia, p. 165) - The borax and the polyvinyl acetate in the glue react to form a polymer.
14. Chemical (S'more Carbon, p. 173) - The sugars in the marshmallow + heat produce CO_2 + H_20; at this temperature these products are gases and escape into the atmosphere. Left behind are some carbon and sugars.
15. Physical (Liquids Are Dense, Too, p. 309) - These two do not really even mix and they can be separated from one another.
16. Physical – It doesn't look the same, but the same molecules are present. A good body shop can make it look the same again.
17. Chemical – An endothermic chemical reaction. With the addition of heat, the liquid part of the egg turns into a solid. This is not a change of state; it is not reversible.
18. Physical – It doesn't look the same and would be really hard to put back together, but you still have glass molecules.
19. Physical (Air Takes Up Space, p. 341) - If you let the air out you have what you started with.
20. Physical and Chemical (Popping Corn, p. 333) - This is both a chemical change and a physical change. Corn pops because of a physical change. Water inside the corn is heated enough that it changes from a liquid state of matter to a gas state. The pressure caused by the expansion needed to go from one state to another results in the corn popping. The outside of the popped corn is still the same thing—it just looks a little different. So this part of the popped corn experienced a physical change. But the popped kernel of corn also experiences a chemical change. The inner part undergoes an endothermic chemical reaction. The proteins inside the corn are irrevocably changed in the presence of the heat source.

NAME ______________________________ DATE ________________

Changes Worksheet: Physical or Chemical?

1.	Painting a piece of wood	physical change	chemical change
2.	Milk goes sour	physical change	chemical change
3.	A match is lit	physical change	chemical change
4.	A sculpture is made from play dough	physical change	chemical change
5.	Food coloring mixed with water	physical change	chemical change
6.	Baby powder is mixed into vinegar	physical change	chemical change
7.	Baking powder is mixed with vinegar	physical change	chemical change
8.	Baking muffins	physical change	chemical change
9.	Painting paper with lemon juice	physical change	chemical change
10.	Making rock candy	physical change	chemical change
11.	The electrolysis of water	physical change	chemical change
12.	Freezing grape juice	physical change	chemical change
13.	Making slime with borax	physical change	chemical change
14.	Cooking a marshmallow at 500°	physical change	chemical change
15.	Mixing oil and water	physical change	chemical change
16.	Running a car into a fence	physical change	chemical change
17.	Frying an egg	physical change	chemical change
18.	Breaking glass	physical change	chemical change
19.	Inflating a balloon	physical change	chemical change
20.	Challenge: Popping corn	physical change	chemical change

Changes Lab #1: Detecting Changes – instructions

CAUTION: This lab requires the use of fire. Students should observe this portion of the lab only from a safe distance.

This lab requires some preparation the day before.

Materials:

- Lab sheet, pencil
- Piece of paper
- Alka-Seltzer tablet
- Water
- Match or lighter
- Two packages of the same flavor of Kool-Aid + ingredients to make Kool-Aid (or any powdered drink mix can be used)
- One glass
- Container to freeze Kool-Aid in
- Pitcher
- Bubble-blowing solution and wand
- Something to burn pieces of paper in (e.g. a cast-iron skillet)
- Scissors

Aloud: Chemical changes are chemical reactions. During a chemical change, the molecules change into different types of molecules. Some signs that there has been a chemical change are gas bubbles forming where there were none before, a solid forming where there was none before, and a change in temperature when you mix the things together.

Physical changes do not change the molecules present. When matter changes the state it is in, that is a physical change. A change of a physical property, such as shape, is a physical change.

Procedure:

The day before:

1. The day before, make about half a packet of Kool-Aid according to the directions and freeze it (ice-cube trays or paper cups work well). You don't need to use all of this packet; just a few ice cubes or one paper-cup full is needed for the lab. Save the other half of the dry powder from this packet and the unopened packet of Kool-Aid for the lab day.

Lab day:

2. The day of the lab, take the frozen Kool-Aid out and let it begin to melt. You will come back to it at the end of the experiment.
3. Have students complete the hypothesis portion of the lab sheet. The hypothesis is a column on the table. Have students keep up with the observation portion of the lab sheet throughout the rest of the lab as you demonstrate each of the following activities.
4. Cut the paper into pieces.
5. Instructor: Put the paper into a nonflammable container and light the pieces.
6. Crush the Alka-Seltzer tablet. Put it in a glass and pour ½ cup of water into the glass.
7. Blow bubbles. The bubbles generated in this experiment DO NOT indicate a chemical change because you still have the same molecules. You have just taken advantage of one of the physical properties of soap solutions.

(continued on the back)

8. Have students look at the dry Kool-Aid powder. Ask what state of matter the Kool-Aid is in, right now. Make the unopened packet of Kool-Aid according to the directions on the package. Put the dry Kool-Aid, melting Kool-Aid, and Kool-Aid in solution next to each other and have them examine all three. If possible, let kids taste all three. There has been a physical change to the Kool-Aid but not a chemical change. Tasting the three samples of Kool-Aid helps kids understand this because they all will taste about the same. The dry Kool-Aid will taste more concentrated.

Instructor's Note:

When answering the hypothesis portion, have students ask themselves:

1.) Do I think there will be the same type of molecules at the end as there were at the start?

2.) Or do I think a new type of molecule was made?

Answers:

	Observations	Results
Cutting paper	Different size and shape but still paper	Physical change
Burning paper	Not paper anymore, now ash and smoke	Chemical change
Alka-Seltzer + water	Made bubbles where there weren't any before	Chemical change
Blowing bubbles	Still have soap; when they pop the liquid, left behind is still soap	Physical change
Adding water to Kool-Aid	Same color, same taste. Now a liquid = change of state	Physical change
Kool-Aid melting	Same color Same taste Now a solid + a liquid	Physical change

Conclusion:

NAME ______________________ DATE ______________

Changes Lab #1: Detecting Changes

In the hypotheses column, write if you think the event will show a physical or chemical change.

In the observations column, note any evidence of new molecules being formed.

In the results column, write if the change was chemical or physical.

	Hypotheses	Observations	Results
Cutting paper			
Burning paper			
Alka-Seltzer + water			
Blowing bubbles			
Adding water to Kool-Aid			
Kool-Aid melting			

Conclusions: (Circle the answer to complete each sentence.)

A physical change creates   new molecules   no new molecules.

A chemical change creates   new molecules   no new molecules.

Changes Demonstration: Chemical Reactions

This activity is a demonstration. As you read the lesson to your students, you will be demonstrating the concepts using Legos. The portion to be read aloud is written in bold and demonstration instructions are written in brackets.

Materials:

- Six Legos—Four of the same size and color (these represent hydrogen atoms) and two more that are the same as each other but that are different from the first four (these represent oxygen atoms).
- Paper, chalkboard, or dry erase board

There is no lab sheet for this lab.

[Before class, make three Lego "molecules." Each Lego molecule has two of the same Legos put together. So there are two H_2 molecules, and one O_2 molecule.]

Chemical reactions are important. They are like little engines making the world change and grow. When you blink, a chemical reaction inside of you made that happen. The warmth from the sun is a result of chemical reactions. Cars turn on because of a chemical reaction. If a car is sitting and rusting, that is also a chemical reaction. When leaves change color, fall off trees, and then grow back again, these are all because of chemical reactions.

During a chemical reaction, bonds between atoms and molecules break and new bonds form. The molecules at the end of the reaction are different from the starting molecules. Not all molecules react with each other. For example, oil and water do not. But some molecules, when they are together, like to make new things. Sometimes they like to play with someone new.

[Show students there are three Lego molecules of two different types while you define the term *reactants* for them.]

In this chemical reaction, H_2 and O_2 are called reactants. The reactants in a chemical reaction are the starting molecules that make new molecules. They are called reactants because H_2 and O_2 like to *react* with each other to make molecules of water, H_2O. To make water, the bonds between the hydrogen atoms and the bonds between the oxygen atoms break.

[Pull apart the three Lego molecules so that you have six Lego atoms.]

New bonds form between the hydrogen atoms and the oxygen atoms, linking them together.

[Now make Lego water. You need two Lego hydrogen atoms stuck on one Lego oxygen atom. This makes two Lego water molecules. Show students the two Lego water molecules while you define products for them.]

The two H_2O molecules made in this reaction are called the products. The product in a chemical reaction is what we call the new molecules that are made.

[Write "reactants → products" on the board.]
[Write the chemical equation, $2H_2 + O_2 = 2H_2O$, on the board.]

(continued on the back)

Just like equations in math, there are equations in chemistry. The chemical equation for making water molecules is $2H_2 + O_2 = 2H_2O$.

[Take the Legos apart and put them back together again, demonstrating how this equation works—start with two molecules of H_2 and one molecule of O_2. Then take the three molecules apart and rearrange them into two water molecules, H_2O.]

The chemical reaction between the gases of hydrogen and oxygen is an explosive reaction. Oxygen and hydrogen really like to make water. When they are put together, they make water with a bang. In fact, this reaction is used to power space rockets because of the amount of energy it provides.

Instructor's Note:

- The purposes of this demonstration are to demonstrate the importance of chemical reactions in our lives, to visually show how chemical reactions create new molecules, and to define reactants and products. A simple example of writing and balancing a chemical equation is also given for the benefit of those students who can grasp this concept now.

Changes Lab #2: Let's Heat Things Up - instructions

Materials:
- Lab sheet, pencil
- Thermometer, science thermometer or kitchen instant-read thermometer
- 1 teaspoon Yeast
- ¼ cup Hydrogen peroxide
- Container you can see through

Aloud: Sometimes when a chemical reaction occurs, heat is released. In fact, any time that you mix things together and heat is released, this is evidence that a chemical reaction has occurred. When this happens, the chemical reaction is called an exothermic reaction. In an exothermic reaction, the temperature at the end of the reaction is higher than the temperature at the beginning. Wood burning is a good example of an exothermic reaction. You can touch the wood before it is lit on fire. Once it is lit, though, you do not want to touch it. The explosive reaction between hydrogen and oxygen is another example of an exothermic reaction. The chemical reaction that happens in today's experiment is an exothermic reaction. It will not get so hot that it burns you or explodes, but you will be able to feel that it has gotten hotter.

Procedure:
1. Complete the hypothesis portion of the lab sheet.
2. Measure the hydrogen peroxide into the clear container. Put the thermometer into the hydrogen peroxide for about three minutes. Have students record the temperature on the lab sheet. Leave the thermometer in the container.
3. Feel the outside of the container. Students can safely put their fingers inside the container to feel the temperature.
4. Sprinkle the yeast into the container and stir it into the hydrogen peroxide. There will be a lot of bubbling. Start monitoring the temperature change right away. It will happen immediately. Wait until the temperature has stopped rising and quickly write down what it is. When the temperature begins to drop, stir the mix a little. The temperature for this experiment never stabilizes at its maximum. Record the temperature.
5. Feel the outside of the cup. Students can safely put their fingers inside the cup to feel the temperature; it will feel warmer. Complete the lab sheet.

Possible Answers:

Hypotheses:
exothermic, greater than

Observations:
These are samples—your answers may vary because starting temperatures will vary.
Starting temperature = 62.6° F = 17° C
Ending temperature = 86.5° F = 30.3° C
Temperature difference = 86.5 – 62.6 = 23.9° F or 30.3 – 17 = 13.3° C

Discussion and Conclusion:
rose, your student's calculation goes here: (23.9) or (13.3); exothermic

Instructor's Notes:
- When hydrogen peroxide molecules are mixed with yeast, the hydrogen peroxide changes into water plus O_2, oxygen gas. This change produces heat. The bubbles you see are oxygen gas escaping into the air. Yeast in this reaction is a catalyst. A catalyst is something that speeds up a chemical reaction.
- To find the temperature difference, always start with the larger of the two numbers, so that the difference is a positive number.
- Make sure the hydrogen peroxide is fresh. Over time, it turns into water and oxygen.

NAME ______________________________ DATE ________________

Changes Lab #2: Let's Heat Things Up

Hypotheses:

I think the chemical reaction of yeast plus hydrogen peroxide is an

____________________ reaction. Because of this, the temperature of the products at the end of the experiment will be (less than/greater than) the temperature of the hydrogen peroxide at the start of the experiment.

Observations:

Starting temperature: [ ]

Ending temperature: [ ]

Use the equation below to find the temperature difference:

Ending Temperature	–	Starting Temperature	=	Temperature Difference
	—		=	

Discussion and Conclusion:

The temperature (rose/fell) by ________ degrees in this experiment.

Therefore, this experiment is a good example of an ____________________ reaction.

Changes Lab #3: Let's Cool Things Down - instructions

Materials:

- Lab sheet, pencil
- Thermometer, science thermometer or kitchen instant-read thermometer
- 1 teaspoon Baking soda
- ¼ cup Lemon juice
- Spoon for stirring
- Container you can see through

Aloud: Some chemical reactions are exothermic reactions, because they release heat, and some chemical reactions are <u>endothermic reactions</u>. Endothermic reactions do not create heat; instead, they absorb heat from their surroundings. In an endothermic reaction, heat is needed to make the reaction happen. In endothermic reactions, the heat needed is often taken from the surrounding air. Baking bread is an endothermic reaction because the bread dough absorbs the heat needed to bake the bread. If the bread dough was not heated, it would not change from dough to bread. For endothermic reactions, the temperature at the end of the reaction will be lower than the temperature at the start of it. This drop in temperature tells you that a chemical change has taken place. When baking bread, the dough gets very hot and absorbs the heat while it is baking. The bread rapidly cools after it is baked. The chemical reaction that happens in today's experiment is an endothermic reaction.

Procedure:

1. Complete the hypothesis portion of the lab sheet.
2. Measure about ¼ cup of lemon juice into the clear container. Put the thermometer into the lemon juice for about three minutes. Record the temperature on the lab sheet. Leave the thermometer in the container. Students can safely put their fingers inside the container to feel the temperature.
3. Sprinkle 1 teaspoon of baking soda into the lemon juice and stir it to help it dissolve.
4. There will be a lot of bubbling. Start monitoring the temperature change right away. It will happen immediately. Wait until the temperature has stopped lowering and quickly write down what it is. When the temperature begins to rise, stir the mix a little. The temperature for this experiment never stabilizes at its minimum. Record the temperature on the lab sheet.
5. Feel the outside of the container. Students can safely put their fingers inside the cup to feel the temperature; it will feel cooler. Complete the lab sheet.

Possible answers:

Hypotheses:
endothermic, less than

Observations:
These are samples—your answers may vary because starting temperatures will vary.
Starting temperature = 72.7° F = 22.6° C
Ending temperature = 63.7° F = 17.6° C
Temperature difference = 72.7 – 63.7 = 9° F = 22.6 – 17.6 = 5° C

Discussion and Conclusion:
fell; your student's calculation goes here: (9) or (5); endothermic

(continued on the back)

Instructor's Notes:

- When baking soda is stirred into lemon juice, energy is needed to break the bonds holding the baking soda together; when this happens, the baking soda dissolves. The energy to break the bonds comes from the surrounding air. When energy is absorbed like that, the temperature of the solution will go down.
- You have done several experiments this year involving endothermic reactions. Most of them involved dissolving salt.
- For an endothermic reaction, the ending temperature is subtracted from the starting temperature—just the opposite of an exothermic reaction. Just remember: To find the temperature difference, always start with the larger of the two numbers, so that the difference is a positive number.

NAME ______________________________ DATE ______________

Changes Lab #3: Let's Cool Things Down

Hypotheses:

I think the chemical reaction of baking soda plus lemon juice is an ____________________ reaction. Because of this, the temperature of the products at the end of the experiment will be (less than/greater than) the temperature of the lemon juice at the start of the experiment.

Observations:

Starting temperature: [ ]

Ending temperature: [ ]

Use the equation below to find the temperature difference:

Starting Temperature – Ending Temperature = Temperature Difference

[ ] – [ ] = [ ]

Discussion and conclusion:

The temperature (rose/fell) by ________ degrees in this experiment.

Therefore, this experiment is a good example of an ______________________ reaction.

NAME ______________________________ DATE ________________

For my notebook

Some Like It Sour, Some Don't

Have you ever tasted a lemon or vinegar? They taste sour. They are both acids, and acids taste sour. The word *acid* comes from the Latin word *acere*, which means sour.

Acid molecules are a special type of molecule. There are many different types of acid molecules, but most have one thing in common. Most acid molecules have hydrogen atoms in the molecule. Not all molecules with hydrogen atoms are acids, though.

When an acid is poured into water, the hydrogen atom comes off of the molecule and floats away. When the hydrogen atom floats away, it leaves its one and only one electron with the rest of the molecule. Hydrogen is such a good sharer. It leaves its electron for its friends to play with. What a nice guy! This type of chemical reaction has a special name. It is called a dissociation reaction. A dissociation reaction happens when a molecule comes apart and leaves or takes electrons with it.

Base molecules are another special type of molecule. You use a base every day. Soap is a base. The detergents your mom uses to wash clothes and clean the house are also bases. Bases feel slippery and do not taste sour; instead, they taste bitter.

Like acids, there are many different types of base molecules. Most bases have one thing in common. Most bases have an oxygen atom *and* a hydrogen atom. When the base is put in water, the oxygen and hydrogen atoms float away from the rest of the molecule. The oxygen and hydrogen atoms stay linked to each other, though. When they float away, they separate and *take* an electron with them. This reaction in action is also an example of a dissociation reaction.

Some people like sour things and some don't. Do you like sour candy? Have you noticed that it's usually the outer coating on sour candy that makes the candy sour? What do you think the people who make sour candy coat it with—an acid or a base? (acid)

Some Like It Sour, Some Don't Lab #1:
Step 1 (or the day you stink everyone out of the house) – instructions

This lab can be completed over two days, or you can wait several hours between parts.

CAUTION: This lab involves boiling water. Only the parent/instructor should handle the boiling water.

Materials:

- One head of **red** cabbage
- Knife for chopping cabbage, optional
- 4 cups Distilled water—if you use tap water, it may affect the acidity or basicity (pH) of the indicator
- Small strainer
- One glass quart jar with a lid (or another container that holds 1 quart)
- Large container with a cover that will hold 4 cups of cabbage plus 4 cups of water
- Four large **white** cone-style coffee filters (at least three coffee filters per person if teaching a group)
- Cookie sheet with a lip around it
- Bowl
- Rubber gloves (if you do not want your fingers dyed purple)
- Gallon-size sealable baggie

There is no lab sheet for Step 1.

Aloud: Today you will make a liquid that changes color if you add an acid or a base to it. It turns red or pink if an acid is mixed into it, and green or blue if a base is mixed into it. Chemicals that change color depending on whether they are mixed with an acid or a base are called <u>indicators.</u> The word *indicate* means "to show." Indicators show if something is an acid, a base, or neither. I should tell you something else about the cabbage indicator you are going to make. It stinks. You will also be making special indicator paper that indicates if a chemical is an acid, base, or neither.

Procedure:

Day 1

1. Instructor: Boil the 4 cups of distilled water.
2. Chop or tear the leaves of the cabbage into small pieces (no larger than 2-inch squares). Chop until you have 4 cups of cabbage. Put the cabbage into the large container. Pour all 4 cups of boiling water over the cabbage. Cover the container. Let the cabbage-water mix cool several hours or overnight.

Day 2, or when the water has cooled

3. Strain the water from the container into the glass jar so that you no longer have any cabbage leaves in the mix. Discard the cabbage leaves. Be careful—the cabbage solution will stain.
4. Pour one cup of cabbage indicator into a bowl. One at a time, begin putting coffee filters into the bowl. When they are completely wet, take the coffee filters out and put them on the cookie sheet to dry. Do not overlap the filters in case they stick to each other. Remember, the cabbage indicator stains. When they are dry, store them in a baggie for future labs.
5. Save the remaining cabbage indicator in a well-sealed container and refrigerate for future labs.

Instructor's Notes:

- This lab will produce several sheets of indicator paper and a few extra cups of cabbage indicator solution that will be used in labs over the next several weeks. **Refrigerate the solution and keep the paper covered and dry.**
- You may want to follow this lab immediately with Step 2 (the next lab) so students can use the cabbage indicator solution right away.

Some Like It Sour, Some Don't Lab #2: Step 2 (or the fun begins) - instructions

Materials:

- Lab Sheet, pencil
- 2 cups Cabbage indicator made during Step 1
- 2 teaspoons Vinegar
- 2 teaspoons Ammonia
- 1 teaspoon Lemon juice
- 1 teaspoon Water
- 1 teaspoon Baking soda
- 1 teaspoon Dish soap
- 1 teaspoon 7-Up or any clear soda
- 1 teaspoon Salt
- Nine small containers or glasses through which students can observe the color changes (if you use plastic containers they could become stained; clear plastic disposable cups work well)
- Newspapers or other material to cover your work surface (to prevent staining)

Aloud: Today we are going to use the cabbage indicator you made in the last lab to assist us in determining if different substances are bases, acids, or neither. The indicator will turn colors, and the colors will tell us if something is a strong or weak acid or base. Acids turn the indicator from purple to pink. The shade of pink depends on the strength of the acid. Bases turn the indicator from purple to green or blue, depending on their strength. Green means it is a strong base, and blue means it is a weaker base. Today you are going to experiment with nine different compounds to see if they are an acid, a base, or <u>neutral</u>. Something is neutral if it is not an acid or a base. Neutral compounds are in the middle of both. If you add a neutral compound to the indicator, the color of the indicator will stay purple.

Procedure:

1. Complete the hypothesis portion of the lab sheet.
2. Pour two cups of cabbage indicator evenly between the nine cups (just eyeball it).
3. Do not put anything except indicator in the first cup. This cup is used as a reference. Starting with the second cup, add the compounds to be tested in the order indicated below (this is the order they are on the lab sheet). Students should record the color of the indicator after each addition on their lab sheets.

   Cup 1 - only cabbage indicator
   Cup 2 - vinegar
   Cup 3 - ammonia
   Cup 4 - lemon juice
   Cup 5 - water
   Cup 6 - baking soda
   Cup 7 - dish soap
   Cup 8 - 7-Up
   Cup 9 - salt

Instructor's Notes:

- Cover the surface where you are conducting this experiment to prevent staining. The cabbage indicator will stain clothes as well.

(continued on the back)

Answers:

Cup	Add	Color	Acid, Base, Neutral
1	nothing	purple	neutral
2	vinegar	pink/red	acid
3	ammonia	green	base
4	lemon juice	pink/red	acid
5	water	purple	neutral
6	baking soda	green	base
7	dish soap	blue	base
8	7-Up	pink	acid
9	salt	purple	neutral

NAME ______________________________ DATE ________________

Some Like It Sour, Some Don't Lab #2: Step 2 (or the fun begins)

Hypotheses: (circle your answer)

I think baking soda will turn the indicator

pink red blue green

I think lemon juice will turn the indicator

pink red blue green

Results/Observations:

Cup	Add	Color	Acid, Base, Neutral
1	nothing		
2	vinegar		
3	ammonia		
4	lemon juice		
5	water		
6	baking soda		
7	dish soap		
8	7-Up		
9	salt		

NAME ______________________________ DATE ____________________

For my notebook

Hydrogen and Oxygen and Hydrogen Make Water

Acids and bases like to play together and make reactions in action happen. When an acid and a base are mixed together, they react with each other, so they are called reactants. The name of this reaction is an <u>acid-base reaction</u>. One of the products of this type of reaction is water; another product is salt. Remember, a product is a molecule made in a chemical reaction. If we write this out as an equation (like you do in math), the equation for an acid-base reaction is: Acid + Base = Water + Salt

Reactants = Products

Acids and bases call this the Let's Make Water Game, and they really like to play it. Acids and bases have the hydrogen and oxygen molecules needed to play this game. Do you remember that water, H_2O, is made from two hydrogen atoms and one oxygen atom? Acid molecules usually have a hydrogen atom. Base molecules usually have an oxygen atom and a hydrogen atom linked together. As you already know, oxygen and hydrogen love to play together. They are best friends, after all!

When acid and base molecules are mixed together, the acid molecules let go of hydrogen atoms, each MISSING ONE electron, and the base molecules let go of oxygen and hydrogen pairs linked together with ONE EXTRA electron. Is it any surprise that acids and bases like to play the Let's Make Water Game? They have everything they need to make water, and you know how much everyone likes water. When they play the Let's Make Water Game, they have a chemical reaction. The name of this type of reaction in action is an acid-base reaction.

Hydrogen and Oxygen and Hydrogen Make Water Lab #1: Let's Make Water – instructions

Perform this experiment over a sink or a plastic tub. You should be prepared for the solution to bubble over and out of the container.

Materials:

- Lab sheet, pencil
- 2 cups White vinegar
- 5 teaspoons of Baking soda
- ¼ cup Cabbage indicator solution
- Tall container that will hold 5 or more cups
- Water
- Indicator paper made in "Some Like It Sour, Some Don't" lab
- Scissors
- Three small cups

Procedure:

1. Cut one of the indicator papers into strips about ½ inch wide. Pour a small amount of water into the first small cup. Mix a small amount of baking soda with a small amount of water in the second small cup. Pour a small amount of vinegar into the third small cup. Quickly dip an indicator paper strip into each of the three cups. Compare the strips for color change and answer questions 1 - 3 on the lab sheet.
2. Pour ¼ cup of indicator solution into the tall container.

(Have students record observations on the chart after each of the following steps.)

3. Stir 1 tablespoon of vinegar into the tall container with the indicator in it. Use an indicator strip to determine if the solution is acidic or basic. Record observations. Record any evidence of a chemical change (bubbling, changing colors).
4. Into the same tall container, stir 1 teaspoon of baking soda into the indicator + vinegar solution. Use an indicator strip to determine if the solution is acidic or basic. Record observations. Record any evidence of a chemical change (bubbling, changing colors).
5. If the solution is basic, stir in 1 teaspoon of vinegar. If the solution is acidic, stir in ¼ teaspoon of baking soda. After you have added the acid or base, use an indicator strip to determine if the solution is acidic or basic. On your lab sheet, circle if you added vinegar, an acid, or baking soda, a base, and record how much you added. Also, record the color the indicator strip turned and any evidence of chemical change. Repeat this two-step process until the indictor paper does not change color from purple. Do not forget to record your results as you go along. Add up the number or teaspoons or tablespoons of vinegar and baking soda it took to reach neutral. Record the number on the lab sheet.
6. Remember, the purpose of each addition is to neutralize the solution. If the solution is acidic (red or pink) add baking soda, which is the base. If the solution is basic (blue or green) add the acid, which is vinegar. Stir and observe the color change before deciding what to add next. As you get close to a neutralized solution, be careful or you can overshoot the neutral point. Your student may want to adjust the amount of vinegar and baking soda added each time, judging for themselves how much is needed to achieve neutral. You might not use all of the rows on the table before you achieve neutral.

Aloud: Water is not an acid and it is not a base. Water is neutral. If you mix an acid with a base, the acid and base molecules will react with each other and make water molecules. If the acid molecules react with the base molecules so that they all turn into water molecules, you will have neutralized the solution. In other words, the solution will not be acidic and it will not be basic; it will be neutral. In fact, this type of acid-base reaction is called a <u>neutralization reaction</u>. You can use your indicator solution and indictor paper to monitor the changes of a solution when going from acidic or basic to neutral. Remember, a change to pink or red indicates the solution is acidic, and a change to blue indicates the solution is basic. If the solution is purple, it is a neutral solution. That means you have made water, water, water!

(continued on the back)

7. Optional Activity: The following activity is not very scientific because there is not any measuring involved. It is a lot of fun, though. Pour more vinegar into the solution until it turns pink. Then pour in baking soda (and stir) until the color of the solution turns blue again. Remember, the baking soda has to dissolve, so be careful with it. You can do this back and forth until you are tired of watching the color change from pink to blue. Be prepared for a lot of bubbling.

Instructor's Notes:

- When you mix an acid and a base together to make a solution neutral, water and salt are created. The solution is neutral because the water is neutral.
- Even though water is produced, making the solution neutral, it's not pure water, so DON'T DRINK IT! It's not poisonous but it won't taste very good.

Possible Answers:

1. basic, blue
2. acidic, red
3. neutral, did not change color

Answers on the table will vary.
Results will vary.

NAME ______________________ DATE ______________

Hydrogen and Oxygen and Hydrogen Make Water Lab #1: Let's Make Water

Observations:

1. Baking soda is acidic / basic / neutral.
   The indicator paper changed to blue / red / it did not change color.
2. Vinegar is acidic / basic / neutral.
   The indicator paper changed to blue / red / it did not change color.
3. Water is acidic / basic / neutral.
   The indicator paper changed to blue / red / it did not change color.

Type of Chemical Added	Amount	Color of Solution	Color of Indicator Paper	A B N (acid, base, or neutral)	Evidence of Chemical Change
Vinegar	1 T				
Baking soda	1 t				
Vinegar or Baking soda					
Vinegar or Baking soda					
Vinegar or Baking soda					
Vinegar or Baking soda					
Vinegar or Baking soda					
Vinegar or Baking soda					
Vinegar or Baking soda					
Vinegar or Baking soda					
Vinegar or Baking soda					

Results:

It took ___________ teaspoons / tablespoons of vinegar and ___________ teaspoons / tablespoons of baking soda to neutralize my solution and make water.

Hydrogen and Oxygen and Hydrogen Make Water
Lab #2: Painting Magic – instructions

Materials:
- Paintbrush
- 1 tablespoon White vinegar in a cup
- 1 tablespoon Water with ½ teaspoon of baking soda mixed into it, in a cup
- Cup of water for rinsing the brush
- Piece of indicator paper
- Protected surface

There is no lab sheet for this activity.

Aloud: If I paint something with a clear liquid, will the color of it change? The answer is yes, of course, if I use the special indicator paper and an acid and a base for paint. Today you are going to write a message or paint a picture by painting on the indicator paper you made. What colors will your picture be?

Procedure:
1. Cover the surface where the students will be painting. Remember the indicator will stain surfaces.
2. Give each student a piece of the indicator paper and a paintbrush.
3. First, have them paint with the vinegar. When they are through, have them paint with the baking soda solution. Make sure they rinse their brush off when going from the acid to the base. When they are through painting with the base, they can go back and forth between the solutions.

Instructor's Notes:
- Students need to be careful they don't saturate the paper. The less "paint" used, the more vivid the colors will be. If they paint over the acid with the base and vice versa, they can change the color back and forth.
- The indicator will begin to fade from the paper with too much painting.

NAME ______________________ DATE ______________

For my notebook

pHunny pHriends

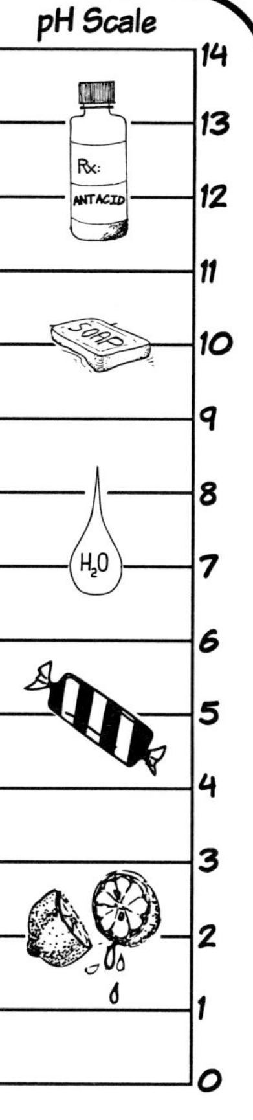

When you look at a thermometer, you are looking at a scale that measures temperature. Scientists have a special scale that measures acids and bases, too. It is called the pH scale. The pH scale is used to tell if something is an acid, a base, or neutral. It also measures the strength of an acid or a base. Can you guess what the H in pH stands for? H stands for hydrogen on the periodic table, and H stands for hydrogen in pH. That's why it is capitalized.

The pH scale goes from 0 to 14. Things that have a pH of 0 are very strong acids. Can you guess what a pH of 14 would indicate? Things with a pH of 14 are very strong bases. Do you know what number is in the middle between 0 and 14? That's right, 7 is the number between 0 and 14. A pH of 7 is at the middle of the pH scale. Water has a pH of 7. Water is said to be neutral. Things that have a pH of 7 are not acids or bases.

The acid side of the pH scale is pH 0 to just before pH 7. The acid side has things like vinegar, the coating that is put on candy to make it sour, and lemon juice. Weak acids have a pH close to pH 7, because they are close to neutral. Strong acids have a pH close to 0.

One important use for the pH scale is to measure the acidity of rain. Normal rain has a pH of 5.6. If rain has a pH of less than 5, it is acid rain. Acid rain is a pollutant. It can kill fish, frogs, and other aquatic animals. It can weaken trees and dissolve stone. If a lake becomes polluted with acid rain, scientists will sometimes add lime, which is a base, to make the lake less acidic.

The base side of the pH scale is from just after pH 7 to pH 14. The base side has things like soap, baking soda, and ammonia. Weak bases have a pH close to pH 7. They are also close to neutral. Strong bases have a pH close to 14. Bases can be useful medicine. Have you ever eaten a lot of beans and experienced painful stomach gas? When you eat foods high in fiber, like vegetables and beans, they can cause your stomach to create too much acid, causing pain, burping, and other interesting reactions. To alleviate this situation, your mom or dad might give you a base called antacid that decreases (or neutralizes) the acid in your stomach and saves the day!

pHunny pHriends Lab #1: Make a pH Scale - instructions

Materials:

- Lab sheet, pencil
- Red, blue, and white watercolor or acrylic paint
- Paintbrush
- Paint pallet or paper plate for blending colors.
- Water for rinsing paintbrush

Aloud: Scientists have a pHunny-looking scale to help them figure out if something is an acid or a base. It is called the pH scale. This scale also measures the strength of an acid or a base.

Procedure - Learn your way around the pH scale:

1. Read the text below aloud as students follow along on the Make a pH Scale lab sheet.

Aloud: Put your finger on the number 7. Water is a 7. If something has a pH of 7, it is not acidic or basic. It is neutral. The number 7 is the midpoint on the scale. If something has a pH below 7, it is acidic. Trace your finger down the pH scale from 7 to 0. As you go down from 7 to 0, things are getting MORE acidic. That means that the strongest acids are closer to 0 than to 7. This part is a little confusing, but that's how it is. Find HCl (hydrochloric acid) at the bottom of the pH scale. HCl is a very strong acid found in your stomach to help digest food.

Trace your finger from 7 to 14 on your worksheet. As you go from 7 to 14, things are getting MORE basic. That means the strongest bases are closer to 14 than to 7. Find NaOH (sodium hydroxide) at the top of the pH scale. NaOH is a very strong base commonly called lye.

Procedure - Paint your pH scale:

Aloud: The indicator paper you made in this unit is also called pH paper. This paper can be used to determine if something is an acid, a base, or neutral. It can also be used to determine the strength of an acid or a base. Just like your indicator paper, on the pH scale, acids are shades of red, and bases are shades of blue. Things that are neutral are very light in color or white. Today you will paint your own pH scale.

2. Paint the pH scale on the worksheet using the watercolor or acrylic paint. Start by placing a small drop of red paint on your pallet. (You will only need a very small amount of red paint.)
3. Load your paintbrush with the red paint, and paint the bottom box on the pH scale (between 0 and 1).
4. Add a small amount of white paint to the red paint on your pallet. Blend to make a slightly lighter shade of red and paint the next box (between 1 and 2). Rinse your brush.
5. Continue this procedure, each time adding a little more white paint to the red on the pallet and painting each consecutive box a lighter and lighter shade of red. STOP AFTER PAINTING THE BOX BETWEEN 5 AND 6 (this box should be painted a light shade of pink).
6. Rinse your brush well and paint the box between 6 and 7 white.
7. Rinse your brush and place a small drop of blue paint on your pallet. (You will only need a very small amount of blue paint.)
8. Load your paintbrush with the blue paint and paint the top box on the pH scale (between 14 and 13).
9. Add a small amount of white paint to the blue paint on your pallet. Blend to make a slightly lighter shade of blue and paint the next box (between 13 and 12). Rinse your brush.
10. Continue this procedure, each time adding a little more white paint to the blue and painting each consecutive box a lighter and lighter shade of blue. Finish with the box between 8 and 7 (this box should be painted a very light shade of blue).

(continued on the back)

Let your pH scale dry before going on with the next part.

Procedure - Using your pH scale:

11. On the back of the worksheet is a list of substances—some basic and some acidic. Back on the front, write the name of each substance on the correct line on the pH scale.
12. On the back of the worksheet, indicate whether each substance is an acid or a base.
13. Answer the questions below the table.

Instructor's Notes:

- You might want to make an extra copy of the worksheet before your student starts painting in case an error is made. It is also a good idea to practice making the shades on scrap paper before painting the original. Shading takes some practice.
- The "H" in pH stands for hydrogen and the "p" stands for potential. A pH scale measures the potential of things to attract hydrogen. pH is a logarithmic scale. The formal definition of pH is the negative logarithm of the hydrogen ion concentration. Because it is a negative logarithm, a 0 value means a high concentration of hydrogen ions and a 14 value means a low concentration of hydrogen ions.

Answers:

baking soda	base
tomato juice	acid
banana	acid
blood	base
soap	base
milk	acid
soda	acid
bleach	base
seawater	base
vinegar	acid
drain cleaner	base
lemon juice	acid
battery acid	acid

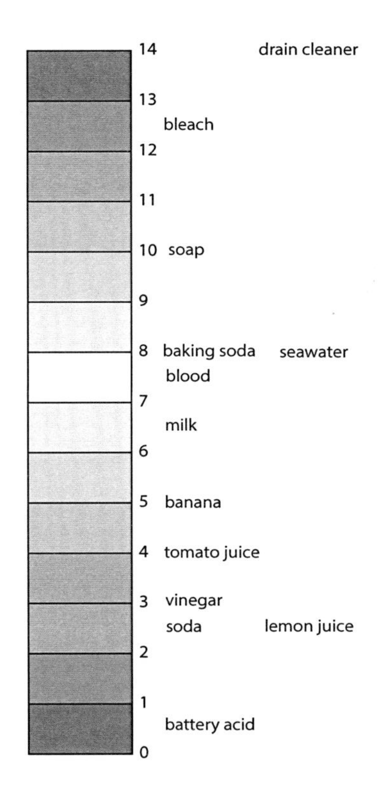

Drain cleaner is the most basic.
Battery acid is the most acidic.
Blood is the weakest base.
Milk is the weakest acid.
Soap is more basic than seawater.
A banana is less acidic than tomato juice.

NAME ______________________ DATE ______________

pHunny pHriends Lab #1: Make a pH Scale - page 1

Bases

Neutral

Acids

14 NaOH (lye) ______

13

______

12

11

10 ______

9

8 ______ ______

______

7

______

6

5 ______

4 ______

3 ______

______ ______

2

1

______

0 HCl (stomach acid)

pHunny pHriends Lab #1: Make a pH Scale - page 2

Item	pH	acid or base?
baking soda	8	
tomato juice	4	
banana	5	
blood	7.5	
soap	10	
milk	6.5	
soda	2.5	
bleach	12.5	
seawater	8	
vinegar	3	
drain cleaner	14	
lemon juice	2.5	
battery acid	.5	

Which substance above is the most basic? ______________________________

Which substance above is the most acidic? ______________________________

Which substance is the weakest base? ______________________________

Which substance is the weakest acid?______________________________

Soap is (more / less) basic than seawater.

A banana is (less / more) acidic than tomato juice.

pHunny pHriends Lab #2:
pHun With Acids – instructions

Materials:

- Lab sheet, pencil
- Cabbage indicator paper
- pH Scale painted in the last lab
- Scissors
- Knife
- Lemon
- Grapefruit
- Lime
- Orange
- Cherry Tomato
- V-8 juice
- 1 tablespoon Distilled water
- Paintbrush
- One small cup
- Glue or tape

Aloud: The cabbage indicator paper and your pH scale can be used to compare the strength of acids. A strong acid, with a low pH, will turn the indicator paper from purple to a vivid pink. A neutral liquid will not change the color of the paper at all. You can tell how strong an acid is by the color of the indicator paper when you put an acid on it. The stronger the acid is, the more pink the paper will be.

Procedure:

1. Complete the hypothesis portion of the lab sheet.
2. Cut seven ½-inch by 2 ½-inch strips of indicator paper. Use one strip of indicator paper for each item.
3. Cut the fruit and cherry tomato in half. You are going to drip one drop of each onto a strip of indicator paper.
4. Use the paintbrush and put a dot of distilled water on one indicator strip—this strip is for reference.
5. Squeeze a drop of juice from the cherry tomato onto a strip of paper. Use the paintbrush to put a dot of V-8 juice onto another strip. Compare these two. The V-8 contains tomato juice and vitamin C—both acids.
6. Squeeze a drop of each citrus fruit onto different strips of indicator paper. Compare the colors resulting from the drips.
7. Glue or tape the strips onto the lab sheet in the order of most acidic to least acidic, placing the water test strip last. Label the strips.
8. Compare your strips to the acid side of pH scale you painted (0 to 7). They should be somewhat similar.

Instructor's Notes:

- The indicator papers will display more vivid colors as they dry.
- The acidity of citrus fruit can vary for each type of fruit. So your answers could be different than mine.
- The indicator you made is only one type of indicator. Different types of indicators turn different colors. There are indicators that turn yellow, those that turn red, and those that become clear, in an acidic solution. There are indicators that turn orange, those that turn blue, and also those that turn yellow, in a basic solution. One type of indicator is litmus. Litmus turns red when the pH is less than 7, and blue when the pH is greater than 7. The pH scale you painted uses a litmus type of color scheme; the acid side of the pH scale is red, and the base side of the pH scale is blue.

(continued on the back)

Possible Answers:

Discussion and Conclusion:

The cherry tomatoes are more acidic than the V-8 juice.
Least acidic fruit = orange   Most acidic fruit = lemon
The lime and the grapefruit will probably tie.

NAME ______________________________ DATE ________________

pHunny pHriends Lab #2: pHun With Acids

Hypotheses:

Which do you think will be more acidic? cherry tomato V-8 juice

List the citrus fruit (lime, lemon, orange, grapefruit) in the order you think they will test, from most acidic to least acidic. (Remember, acidic things are sour.)

1. ________________ most acidic
2. ________________
3. ________________
4. ________________ least acidic

Results: (Glue and label your indicator strips in the spaces below.)

Glue most acidic here (darkest pink)	________________
	________________
	________________
	________________
	________________
Glue least acidic here (lightest pink)	________________
Glue water strip here	Water

Discussion and Conclusion:

Which was more acidic: cherry tomatoes or V-8? ________________

Which citrus fruit was the most acidic? ________________

Which citrus fruit was the least acidic? ________________

Did any seem to be tied for acidity? If yes, which ones?

____

NAME ______________________________ DATE ________________

For my notebook

Building Teeth

Do you remember losing your first tooth? Did you wonder what type of chemical reaction your body used to make a new one? Probably not, because you didn't know about chemistry back then. How about your bones? They are hard solids, and they grow with the rest of your body. Calcium helps your bones and teeth grow strong. You get calcium into your body by eating foods like milk, yogurt, salmon, and broccoli. The calcium dissolves in your stomach and enters your blood as a liquid. What kind of chemical reaction does your body use to take calcium out of a liquid solution and turn it into teeth and bones, both solids? These are the types of questions about chemical reactions that chemists ask.

The reaction in action that makes teeth and bones grow is called a precipitation reaction. Most liquids have molecules dissolved in them. In a precipitation reaction, these liquids with dissolved molecules, mix together to make new molecules that are solids. Your body uses precipitate reactions to help turn the calcium in a liquid into solid bones and teeth.

Precipitation reactions are important in your body and in nature. Your teeth and bones are made using precipitation reactions. Coral reefs are formed through precipitation reactions; seashells and snail shells are, too.

Building Teeth Lab: Precipitates – instructions

This is a two-day lab.

Materials:
- Lab sheet, pencil
- Sealable baggie
- Six pieces of chalk
- ⅔ cup White vinegar
- Pitcher with a pour spout that will hold the vinegar and chalk solution
- Plastic wrap
- Clear glass container, that holds 2 cups
- Container for mixing water and baking soda
- ½ cup Warm water
- 3 teaspoons Baking soda

Day 1

Aloud: We cannot get into someone's body to watch teeth and bones grow, but we can make a precipitate that has calcium in it. A precipitate is what the solid is called that forms in a precipitation reaction. Chalk has calcium in it. It is made up of molecules called calcium carbonate. This is not the same type of calcium that makes up your teeth and bones. Calcium carbonate is an important part of seashells and coral reefs, though. Today you will dissolve calcium carbonate in vinegar. Tomorrow, you will precipitate the calcium carbonate back into its solid form. What will the solid do if it is less dense than the liquid? What will it do if it is more dense than the liquid?

Procedure:
1. Put the chalk pieces into a baggie, and crush them into very small pieces and chalk dust.
2. Pour the vinegar and chalk into a clear glass container. Stir the mixture a few times. Cover with plastic wrap and leave the mixture to sit overnight.
3. Answer the first two questions on the lab sheet.

Day 2

Aloud: Today you are going to mix two liquids and make a solid form. When molecules in liquids combine to make a solid, the solid is called a precipitate. The first liquid has the calcium carbonate that was dissolved into the vinegar overnight. The second liquid has baking soda dissolved in it. While we're are at it, we will check out the density of the solid precipitate that forms.

Procedure:
1. Use a spoon or a spatula and skim any foam from the top surface of the chalk + vinegar solution. Do not stir up the liquid.
2. Pour the clear liquid from the chalk + vinegar solution into a clear glass container that holds 2 cups. This liquid is the dissolved calcium carbonate. Be careful when you pour the liquid; do not pour any of the solid that is sitting on the bottom of the container.
3. In another glass, mix ½ cup of warm water with 3 teaspoons of baking soda. Stir this until most of the baking soda has dissolved.
4. Pour the baking soda solution into the glass container with the dissolved calcium carbonate. Be careful not to pour any undissolved baking soda into the calcium carbonate solution.

(continued on the back)

5. Observe as calcium carbonate precipitates back out of the solution. The solution will be very cloudy at first. If you observe carefully, it looks like snow is falling in the glass. Wait about 20 minutes for the precipitate to fall to the bottom.
6. Pour out the liquid from the precipitate mix. Check out the precipitate in the bottom of the glass. It is safe to touch.
7. Complete the lab sheet.

Instructor's Notes:

- The calcium in people's teeth and bones is calcium phosphate, $Ca_3(PO_4)_2$, not calcium carbonate. Calcium carbonate is a common calcium supplement. When someone takes calcium carbonate supplements, his body dissolves the calcium carbonate. This dissolved calcium carbonate is eventually precipitated out of the solution in the form of $Ca_3(PO_4)_2$, making and strengthening teeth and bones.
- The molecular formula for baking soda is $NaHCO_3$.

Possible Answers:

White, soft, chalky feeling, dry

When chalk and vinegar are mixed, bubbles form, indicating a chemical reaction.

The first picture should show a cloudy solution. The second picture should show a clear-like solution with a white layer on the bottom of the glass.

The precipitate is more dense—you know this because it sinks to the bottom.

White, soft, chalky feeling, white, smells faintly of vinegar

The formation of a precipitate is evidence of a chemical change.

NAME ______________________ DATE ______________

Building Teeth Lab: Precipitates

Describe the chalk dust before it is mixed with vinegar.

A chemical change occurred when chalk and vinegar were mixed together. What is the evidence supporting this statement?

Observations:

My solution immediately after mixing them together.

My solution 20 minutes later.

Which is more dense: the liquid or the precipitate? How do you know?

Write a description of the precipitate formed in the reaction.

A chemical change occurred when the solutions with calcium carbonate and the baking soda were mixed together. What is the evidence supporting this statement?

NAME ______________________ DATE ______________

For my notebook

Combustion Action

A burning candle is evidence of a chemical reaction in action. Fires burning, exploding fireworks, and cars turning on and going are all examples of combustion reactions. Almost anytime anything burns, it is a combustion reaction. Combustion reactions release heat, so they are exothermic.

ALL combustion reactions have one thing in common. They all react or respond to oxygen gas. Do you remember the little O_2 molecule in the air? In order for things to burn, there needs to be O_2 molecules present. The source of oxygen is usually the air around us. That is why people are supposed to smother a campfire with dirt when they leave a campsite. Even if there are hot coals in the ashes, if the oxygen in the air cannot get to the coals, the fire will die out. That is because as long as the fire burns, oxygen gas is being used up. Eventually there is not enough oxygen gas for the fire to keep burning. When this happens, the fire will go out.

In general, every combustion reaction starts with something that burns plus oxygen. These are called the reactants. The products or results of a combustion reaction are some new molecules and heat.

Something that burns + O_2 ⟶ New molecules + Heat

Reactants ⟶ Products

An Example of a Combustion Reaction:

Wood + O_2 ⟶ Ash + Smoke + Heat

Reactants ⟶ Products

Combustion Action Lab #1: Playing with Fire - instructions

CAUTION: This lab involves the use of matches and fire. Only adults should light or handle the candles.

Materials:

- Lab sheet, pencil
- Three different-size glass jars (The greater the difference, the better. But all the jars need to be able to fit over the candles)
- Roasting pan, large enough to fit all three jars at the same time
- Sand (enough to fill the pan to a 1-inch depth)
- Three votive candles, all the same size
- Matches or a lighter
- Stopwatch or watch with a second hand

Aloud: Oxygen and fire are a team. Without oxygen, you cannot have fire. Fire needs oxygen to keep burning. In this lab, three candles will be lit, and different-size jars will be placed over the top of them. Do you think the size of each jar will affect how long each candle stays lit?

Procedure:

1. Complete the hypothesis portion of the lab sheet.
2. Fill the roasting pan with sand so that the depth is 1 inch. Put the candles in the sand in a row. Make sure the jars will fit over each candle and into the sand. The sand is there to make sure no oxygen can creep into the jars.
3. Instructor: Light the candles. Wait a short time to make sure they are well lit.
4. Instructor: Put the jars over the burning candles at the same time (you will need extra adult hands for this). [Alternatively, you can place the jars one at a time over the candles, timing each one individually.]
5. Start the timer right away after you put the jars on the candles. Record the time it took for each candle to go out. It might help to have one person operating the timer and one person recording the time. Do not turn the timer off until the last candle goes out.
6. Record the results on the lab sheet.

Instructor's Notes:

- You can use water in place of sand in the roasting pan. If you use water, make sure the candles are secured to something (like clay) so that they cannot fall over. If the candles get wet, they will not work in this experiment.
- You might need to cut the candles off at the bottom end to make the jars fit over them. There has to be some head space for the candles.

Possible Answers:

Hypotheses:

I think the candle covered by the large jar will stay lit the longest.
I think the candle covered by the small jar will go out first.

Observations:

The drawing of the jars should emphasize the different sizes.

Results:

The timed results will vary.

Discussion and Conclusion:

- The candle burned for the longest amount of time when covered with the large jar. The candle burned for the medium amount of time when covered by the medium jar. The candle burned for the smallest amount of time when covered by the small jar.
- This experiment shows that if all the air around a fire is used up, the fire will go out. That is because in a combustion reaction, you need oxygen from the air to keep the fire going; it is an essential reactant.

NAME ______________________________ DATE ____________________

Combustion Action Lab #1: Playing with Fire

Hypotheses:

I think the candle covered by the jar will stay lit the longest. large medium small

I think the candle covered by the jar will go out first. large medium small

Observations:

My jars looked like this over the candles:

Results:

	Small Jar	Medium Jar	Large Jar
Candle went out at			

Discussion and Conclusion:

In which jar did the candle stay burning the longest?

In which jar did the candle burn the least amount of time?

This experiment shows that if all the air around a fire is used up, the fire will

____________________________ . That is because in a combustion reaction you

need ______________________ from the air to keep the fire going.

Combustion Action Lab #2: Burning Money– instructions

CAUTION: This lab involves the use of fire and burning objects. Students should observe this lab only from a safe distance.

Materials:

- Dollar bill
- Tongs
- Lighter
- ¼ teaspoon Salt
- ½ cup 91% Rubbing alcohol
- ½ cup Water
- Pan
- Measuring cup
- Sink or other nonflammable surface

There is no lab sheet for this lab.

Aloud: The lab for today is an exciting combustion reaction. You are going to watch as a dollar bill is lit on fire. There will be a whole roomful of oxygen around, so today the oxygen will not be used up in the experiment. Do you think this chemical reaction is endothermic or exothermic? (exothermic)

Procedure:

1. Pour the rubbing alcohol, water, and salt into the pan, then stir this to dissolve and mix in the salt. Put the dollar bill into the pan and let it become saturated. Be careful not to get any alcohol on you. DO NOT LIGHT THE DOLLAR BILL WITH ALCOHOL ON YOUR HANDS.
2. Use the tongs to take out the bill. Hold the bill over the pan until the liquid has stopped draining from it. Do not let the bill dry.
3. Move the still-damp bill far away from the pan and over a sink or other fireproof area. If you get any alcohol on your hands, wash your hands NOW. DO NOT LIGHT THE DOLLAR BILL WITH ALCOHOL ON YOUR HANDS.
4. Keep holding the bill with the tongs. Light the bill with the lighter. Let the bill keep burning until the fire goes out. When the flame goes out, light the lighter again and run it along the bill. The bill will not relight. Have everyone check the bill out so students can see that it did not burn.

Aloud: Before the bill was lit, it was soaked in rubbing alcohol and water. Rubbing alcohol burns at a lower temperature than water. The fire that you saw came from the combustion of the rubbing alcohol, not the dollar bill. As the fire burned, the rubbing alcohol on the bill burned off the bill. The reaction in action turned the rubbing alcohol into carbon dioxide and water molecules. When the rubbing alcohol burned off the dollar bill, the water still on the bill put the fire out!

Instructor's Notes:

- Do not use fake money. Real bills are not paper; they are similar to cloth that is woven. That is one reason they do not burn in this experiment.
- This experiment creates a similar situation to that created when food is flambéed.
- The specific chemical reaction for this experiment is:

$$C_2H_5OH + 4O_2 \longrightarrow 2CO_2 + 3H_2O + \text{heat}$$

something to burn (rubbing alcohol) + oxygen ⟶ new molecules (carbon dioxide & water) + heat

reactants ⟶ products

Reactions in Action - Crossword Vocabulary Review

Across

1. The new molecules made in a chemical reaction.
6. Heat is absorbed in this type of chemical reaction.
8. In this type of change, the appearances changes, but the molecules do not.
9. A chemical with a pH of 12 is this.
10. A chemical that shows if something is an acid, a base, or neither.
11. The starting molecules in a chemical reaction.

Down

2. Combination, separation, or rearrangement of atoms into new and different molecules. (Two words)
3. Heat is released in this type of chemical reaction.
4. Oxygen is a necessary reactant for this type of chemical reaction.
5. A chemical with a pH of 7 is this.
7. A chemical with a pH of 1 is this.
8. Measures the acidity and basicity of solutions, with values from 0 to 14. (Two words)

Crossword Puzzle Answer Keys

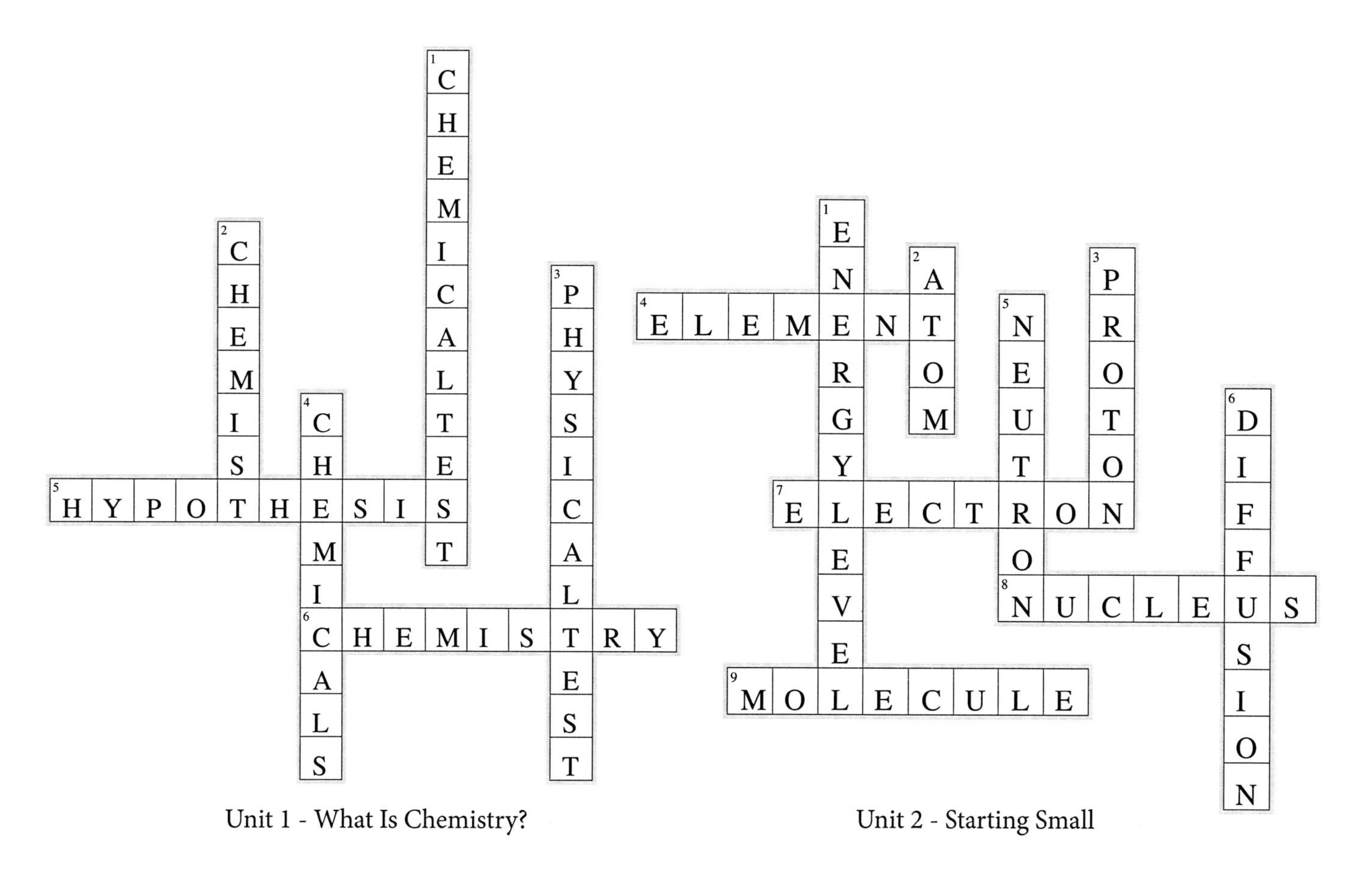

Unit 1 - What Is Chemistry?

Unit 2 - Starting Small

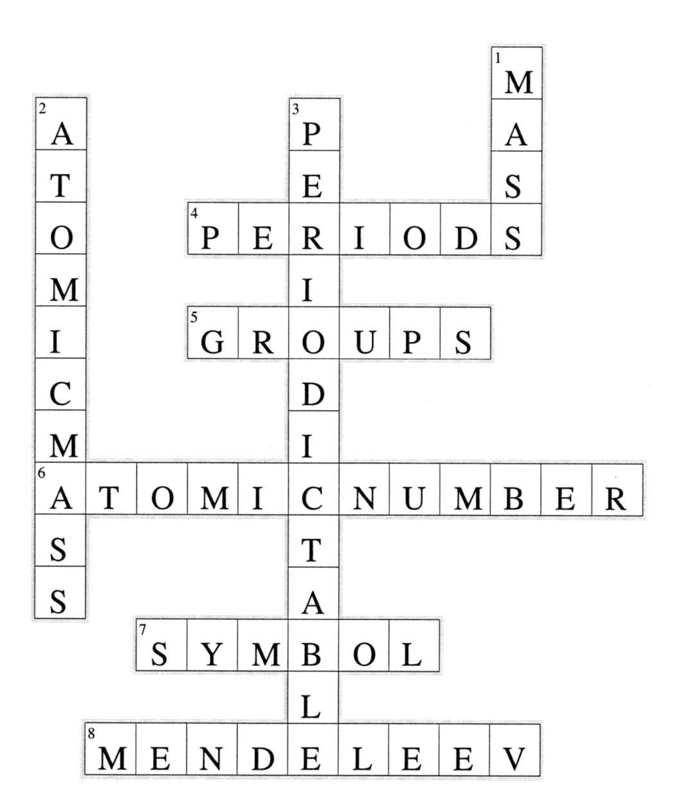

Unit 3 - The Chemist's Alphabet Defined

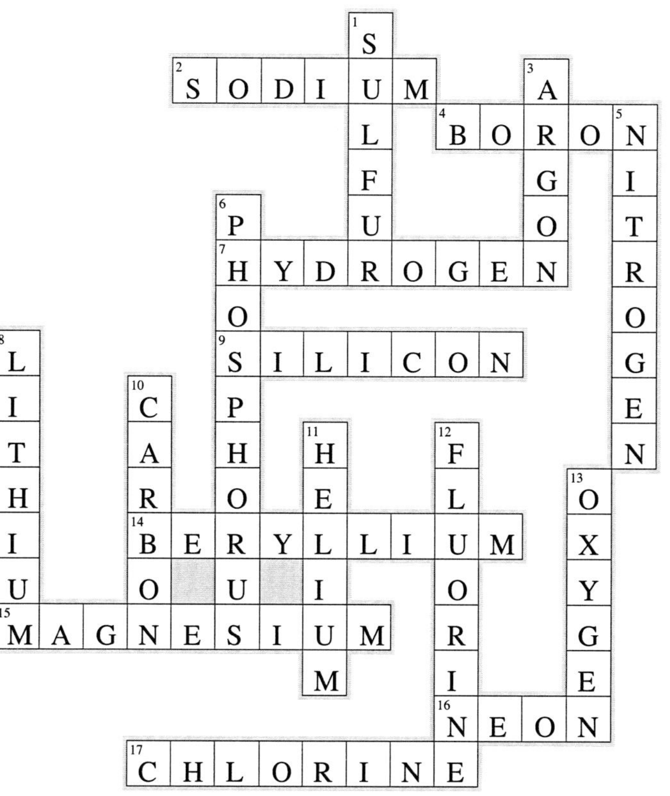

Unit 4 - The Chemist's Alphabet Applied

Bonus: Aluminum, Al

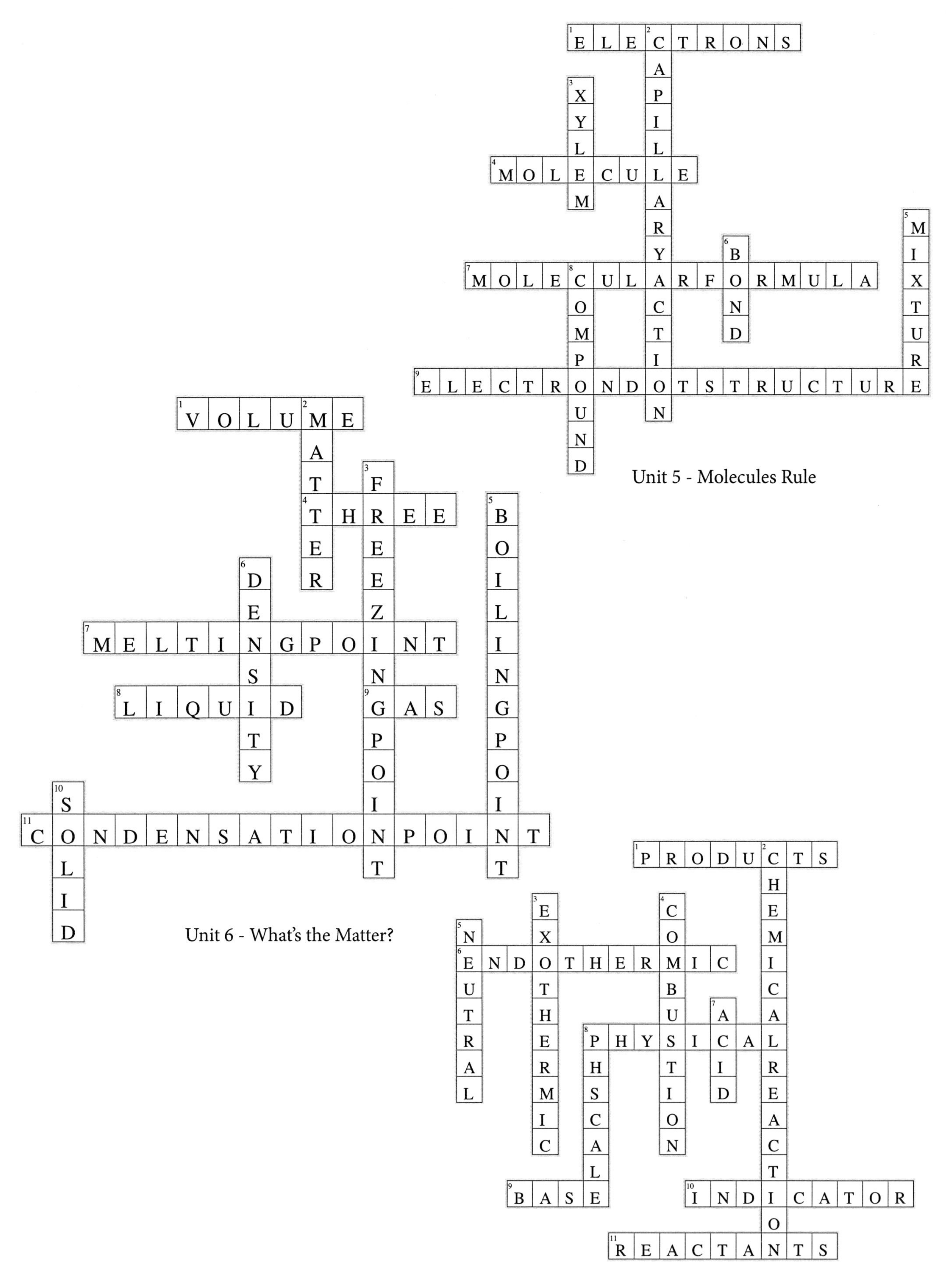

Unit 5 - Molecules Rule

Unit 6 - What's the Matter?

Unit 7 - Reactions in Action

Vocabulary

Acid – From the Latin word *acere*, meaning "sour." Acids are a special type of molecule. Most acids have a hydrogen atom that dissociates from the molecule without its electron when it is put in water. Acidic solutions have a pH below 7.

Acid-Base Reaction – A type of chemical reaction with the following form: acid + base = water + salt.

Acid Rain – Rain with a pH of less than 5. Acid rain is a pollutant that can kill aquatic animals and amphibians, weaken trees, and dissolve stone.

Acrostic – A type of poem. A letter in each word of the poem spells out another message.

Aluminum (Al) – Aluminum is in Group 13 in the 3rd period. Its atomic number is 13, and its atomic mass is 27. Aluminum is the most common metal on Earth's surface. It is the third most common element in Earth's crust. It is a strong, light, and corrosion-resistant metal.

Ammonia – The chemical formula for ammonia is NH_3. Ammonia is used to make fertilizers, explosives, and cleaning products.

Anode – A positively charged electrode.

Antacid – A substance that neutralizes acid.

Argon (Ar) – Argon is in Group 18 in the 3rd period. Its atomic number is 18 and its atomic mass is 40. Argon is the third most common gas in Earth's atmosphere. It is used to fill lightbulbs.

Atmosphere – All the gas surrounding a planet.

Atom – The basic building blocks of all matter. Atoms are made from protons, electrons, and neutrons.

Atomic Mass Unit – The unit of measurement used to tell the mass of an atom—abbreviated as a.m.u.

Atomic Number – Equals the number of protons in the nucleus of an atom. Each element has its own unique atomic number. This number can be used to identify an element and is found in the periodic table.

Atomic Theory of Matter – John Dalton proposed this theory in 1803. The theory states that 1.) All things are made of atoms. 2.) Atoms of the same element are the same, and atoms of different elements are different from each other. 3.) Compounds are formed when different types of atoms combine with each other.

Base – A special type of molecule. Most bases have an oxygen-hydrogen pair that dissociates from the molecule with an extra electron when put in water. Basic solutions have a pH above 7.

Beryllium (Be) – Beryllium is in Group 2 in the 2nd period. Its atomic number is 4, and its atomic mass is 9. Beryllium is found in emeralds. It is a deadly poison if eaten.

Boiling Point – The point in temperature at which a compound goes from a liquid to a gas.

Bonds – The places where atoms link together to make molecules.

Boron (B) – Boron is in Group 13 in the 2nd period. Its atomic number is 5, and its atomic mass is 11. It is used to make the green color in fireworks.

Buoyancy – The ability of things to float.

Calcium Carbonate – The chemical formula for calcium carbonate is $CaCO_3$. Calcium carbonate is found in rocks. It is an important molecule in the shells of marine animals, snails, and birds.

Capillary Action – The ability of a substance to draw another substance into it.

Carbon (C) – Carbon is in Group 14 in the 2nd period. Its atomic number is 6, and its atomic mass is 12. It has two elemental forms: graphite and diamond. All living things on Earth have carbon in them. It is the sixth most common element in the universe.

Carbon Dioxide – The chemical formula for carbon dioxide is CO_2. Carbon dioxide is a gas found in Earth's atmosphere.

Catalase – A compound found in your blood and in many other living things as well.

Catalyst – Something that speeds up a chemical reaction.

Cathode – A negatively charged electrode.

Cells – The basic units of all living things.

Cellulose – A special type of molecule found in plants.

Chemical Change – Rearrangement of atoms such that new and different molecules are made. A chemical change is necessary for a chemical reaction to occur.

Chemical Family or Group – This is a column of the periodic table. There are 18 groups.

Chemical Reaction – A combination, separation, or rearrangement of atoms into new and different molecules.

Chemical Tests – These are tests performed on compounds to see how they behave chemically.

Chemicals – All matter is made of chemicals. Chemicals are made up of atoms and molecules.

Chemist – A person who studies chemicals and the matter they make.

Chemistry – The science studying chemicals and the matter they make.

Chlorine (Cl) – Chlorine is in Group 17 in the 3rd period. Its atomic number is 17, and its atomic mass is 35. It is found in bleach and table salt.

Chlorophyll – The molecule that makes plants green.

Colloid – A solid suspended in a liquid. It has properties between a solid and a liquid.

Combustion Reactions – A type of chemical reaction with the following form, something that burns + O_2 → new molecules + heat. Combustion reactions are exothermic.

Compound – A group of molecules that is all the same kind.

Condensation Point – The point in temperature at which a compound goes from a gas to a liquid.

Crystals – Solids that form in an orderly, repeating pattern, e.g. sugar and NaCl.

Dalton, John – The famous English chemist (1766–1844) who proposed the Atomic Theory of Matter.

Democritus – The Greek who 2,400 years ago said that everything is made of particles called atoms.

Dense – Things that are dense have molecules very close together, as in a solid.

Density – The physical property that measures the amount of stuff in a given space.

Diffusion – A transport mechanism, where molecules move from an area of high concentration to an area of low concentration.

Dissociation Reaction – Occurs when a molecule comes apart and leaves or takes electrons.

Electric Charge – An amount of electrical energy.

Electrolysis – A method of separating bonded atoms using an electrical current.

Electrolyte – A negatively or positively charged particle (ion) that conducts electricity.

Electron Dot Method – This is a method used for drawing atoms and molecules. Using this method, the electrons in the outer energy level are drawn around the elemental symbol.

Electron Dot Structures – This is the name used for the atoms and molecules drawn using the Electron Dot Method.

Electrons – Small particles that orbit around the nucleus of an atom. Electrons have a negative charge.

Element – Matter that contains a specified type of atom; all of which all have the same number of protons in the nucleus. There are 118 different types of elements.

Endothermic Reactions – A type of reaction where the temperature at the end of the reaction is lower than the temperature at the beginning. Heat is absorbed in this type of chemical reaction.

Energy Level – Where the electrons of an atom are found.

Exothermic Reaction – A reaction where the temperature at the end of the reaction is higher than the temperature at the beginning. Heat is released in this type of chemical reaction.

Fertilizer – A chemical added to soil as a food for plants.

Fluid – The ability to flow.

Fluoride – Used in toothpaste to help strengthen teeth.

Fluorine (F) – Fluorine is in Group 17 in the 2nd period. Its atomic number is 9, and its atomic mass is 19. It is a very reactive element. Fluorine is found in fluoride toothpaste to help strengthen teeth.

Freezing Point – The point in temperature at which a compound goes from a liquid to a solid.

Gas – A state of matter. In a gas the molecules have a lot of space between them. Gases do not have definite volume or a definite shape.

Gold (Au) – Gold is in Group 11 in the 6th period. Its atomic number is 79, and its atomic mass is 197. It is a metal.

Helium (He) – Helium is in Group 18 in the 1st period. Its atomic number is 2, and its atomic mass is 4. The second most common element in the universe and the sixth most common gas in Earth's atmosphere.

Hydrochloric Acid – The chemical formula for hydrochloric acid is HCl. Hydrochloric acid is a very strong, corrosive acid with a low pH that has many industrial uses and is found in gastric juices.

Hydrogen (H) – Hydrogen is in Group 1 in the 1st period. Its atomic number is 1, and its atomic mas is 1. It is the smallest atom, the most common element in the universe, and the tenth most common element in Earth's crust.

Hydrogen Bonding – This type of bond can occur between molecules, where one of the molecules has hydrogen as a part of it. Hydrogen bonds hold water molecules together.

Hydrogen Peroxide – The chemical formula for hydrogen peroxide is H_2O_2. It is used as an oxidizing and bleaching agent, and for antiseptic uses.

Hypothesis – A reasoned proposal predicting an outcome of an experiment.

Indicators – Chemicals that show if something is an acid, a base, or neither.

Iron (Fe) – Iron is in Group 8 in the 4th period. Its atomic number is 26, and its atomic mass is 56. It is a metal.

Lead (Pb) – Lead is in Group 14 in the 6th period. Its atomic number is 82, and its atomic mass is 207. It is a metal.

Liquid – A state of matter. Liquids have a definite volume but do not have a definite shape.

Lithium (Li) – Lithium is in Group 1 in the 2nd period. Its atomic number is 3, and its atomic mass is 7. The name *lithium* comes from the Greek word for "stone."

Magnesium (Mg) – Magnesium is in Group 2 in the 3rd period. Its atomic number is 12, and its atomic mass is 24. It is the eighth most common element in the universe and the seventh most common element in Earth's crust. It is found in Epsom salt and seawater.

Mass – A measurement for how much matter is in an object. Unlike weight, mass is not affected by gravity.

Matter – All things that are made of atoms and molecules.

Melting Point – The point in temperature at which a compound goes from a solid to a liquid.

Mendeleev, Dmitri – A famous Russian scientist (1834–1907) who created the periodic table.

Mixture – A group of different kinds of molecules.

Molecular Formula – Formulas that indicate the amount and type of atoms present in a molecule, using elemental symbols and numerical subscript.

Molecules – A group of two or more atoms bonded together.

Neon (Ne) – Neon is in Group 18 in the 2nd period its atomic number is 10, and its atomic mass is 20. It is the fourth most common element in the universe and the fifth most common gas in Earth's atmosphere. It is used to make neon lights.

Neutral – Not an acid or a base.

Neutralization Reaction – Acid-base reactions perfectly calibrated to result in a neutral solution—pH = 7.

Neutrons – Particles in the nucleus of an atom. Neutrons have no charge.

Nitrogen (N) – Nitrogen is in Group 15 in the 2nd period. Its atomic number is 7, and its atomic mass is 14. It is the fifth most common element in the universe. 78% of Earth's atmosphere is nitrogen, making it the most common element in Earth's atmosphere.

Nucleus – The nucleus is the center of an atom, containing the protons and neutrons.

Oxygen (O) – Oxygen is in Group 16 in the 2nd period. Its atomic number is 8, and its atomic mass is 16. It is the third most common element in the universe, the second most common gas in Earth's atmosphere, and the most common element in Earth's crust and in the ocean.

Periodic Table of the Elements – The main reference document of chemistry. The periodic table tells the properties of the elements.

Periods – The rows of the periodic table. There are seven periods. The period an element is in tells you the number of energy levels that element has.

pH Scale – A scale that measures the acidity and basicity of solutions. The pH scale ranges from 0 (strong acid) to 7 (neutral) to 14 (strong base).

Phosphorescent – The emission of light—the property of glowing in the dark.

Phosphorus (P) – Phosphorus is in Group 15 in the 3rd period. Its atomic number is 15, and its atomic mass is 31. It is used to make fireworks.

Physical Change – This is a change in appearance of something without there being a change in the molecules present.

Physical Tests – These observations and tests are performed to determine the physical properties of compounds.

Polymer – A long chain of molecules that stretch and bend.

Potassium (K) – Potassium is in Group 1 in the 4th period. Its atomic number is 19, and its atomic mass is 39. It is the seventh most common element in Earth's crust.

Precipitation Reaction – A special type of chemical reaction where liquids with dissolved molecules mix together and make new molecules that are solids.

Precipitate – The solid that forms in a precipitation reaction.

Products – The new molecules made in a chemical reaction.

Protons – Particles in the nucleus of an atom. Protons have a positive charge.

Reactants – The starting molecules in a chemical reaction.

Scanning-Tunneling Microscope – A special type of microscope used to see atoms.

Silicon (Si) – Silicon is in Group 14 in the 3rd period. Its atomic number is 14, and its atomic mass is 28. It is the seventh most common element in the universe and the second most common gas in Earth's crust. It is found in sand.

Silver (Ag) – Silver is in Group 11 in the 5th period. Its atomic number is 47, and its atomic mass is 108. It is a metal.

Sodium (Na) – Sodium is in Group 1 in the 3rd period. Its atomic number is 11, and its atomic mass is 23. It is the sixth most common element in Earth's crust. It is found in table salt.

Sodium Chloride – The chemical formula for sodium chloride is NaCl. It is common table salt.

Sodium Hydroxide – The chemical formula for sodium hydroxide is NaOH. It is a strong base with a pH of 14. The common name is lye. It is used in making paper, drinking water, and detergents.

Sodium Hypochlorite – The chemical formula for sodium hydroxide is NaClO. Its common name is bleach.

Solid – A state of matter. Solids have a definite volume and a definite shape.

Solute – A substance dissolved in another substance, called the solvent.

Solution – A mixture with two parts to it, the solvent and the solute.

Solvent – A substance in which another substance, the solute, is dissolved.

States of Matter – The three forms of matter—solid, liquid, and gas.

Steam – The name for water when it is a gas.

Subscript – A word, number, or symbol that appears below the text—for example the 2 in H_2O is in subscript notation.

Sulfur (S) – Sulfur is in Group 16 in the 3rd period. Its atomic number is 16, and its atomic mass is 35. It is found in eggs, volcanoes, fireworks, and matches.

Sulfuric Acid – The chemical formula for sulfuric acid is H_2SO_4. It is a strong acid.

Symbols – In chemistry, the abbreviations for the names of the elements.

Theory – A statement of what are held to be the general laws, principles, or causes of something known or observed. In science, theories are ideas that interpret facts and explain how or why something works.

Universal Solvent – Water. It is the universal solvent, because it is the main solvent into which solutes are dissolved.

Volume – A physical property that measures the amount of space something takes up.

Xylem – Long, straw-like tubes in plants made of cellulose molecules. Xylem are used for water transport.

Zinc (Zn) – Zinc is in Group 12 in the 4th period. Its atomic number is 30, and its atomic mass is 65. It is a metal.

Index

Page locators that appear in **bold** text refer to the glossary entry, and page locators that are underlined refer to the definition in the text.

C

D

E

F

Pandia PRESS

Pandia PRESS

Pandia PRESS

T

U

V

W

X

Z